Symmetry in Electromagnetism

Symmetry in Electromagnetism

Editors

Albert Ferrando
Miguel Ángel García-March

MDPI • Basel • Beijing • Wuhan • Barcelona • Belgrade • Manchester • Tokyo • Cluj • Tianjin

Editors
Albert Ferrando
University of Valencia
Spain

MiguelÁngel García-March
Mediterranean Technology Park
Spain

Editorial Office
MDPI
St. Alban-Anlage 66
4052 Basel, Switzerland

This is a reprint of articles from the Special Issue published online in the open access journal *Symmetry* (ISSN 2073-8994) (available at: https://www.mdpi.com/journal/symmetry/special_issues/symmetry_electromagnetism).

For citation purposes, cite each article independently as indicated on the article page online and as indicated below:

LastName, A.A.; LastName, B.B.; LastName, C.C. Article Title. *Journal Name* **Year**, *Article Number*, Page Range.

ISBN 978-3-03943-124-3 (Hbk)
ISBN 978-3-03943-125-0 (PDF)

© 2020 by the authors. Articles in this book are Open Access and distributed under the Creative Commons Attribution (CC BY) license, which allows users to download, copy and build upon published articles, as long as the author and publisher are properly credited, which ensures maximum dissemination and a wider impact of our publications.

The book as a whole is distributed by MDPI under the terms and conditions of the Creative Commons license CC BY-NC-ND.

Contents

About the Editors vii

Preface to "Symmetry in Electromagnetism" ix

Albert Ferrando and Miguel Ángel García-March
Symmetry in Electromagnetism
Reprinted from: *Symmetry* **2020**, *12*, 685, doi:10.3390/sym12050685 1

Manuel Arrayás and José L. Trueba
Spin-Orbital Momentum Decomposition and Helicity Exchange in a Set of Non-Null Knotted Electromagnetic Fields
Reprinted from: *Symmetry* **2018**, *10*, 88, doi:10.3390/sym10040088 5

Manuel Arrayás, Alfredo Tiemblo and José L. Trueba
Null Electromagnetic Fields from Dilatation and Rotation Transformations of the Hopfion
Reprinted from: *Symmetry* **2019**, *11*, 1105, doi:10.3390/sym11091105 21

Francisco Mesa, Raúl Rodríguez-Berral and Francisco Medina
On the Computation of the Dispersion Diagram of SymmetricOne-Dimensionally Periodic Structures
Reprinted from: *Symmetry* **2018**, *10*, 307, doi:10.3390/sym10080307 39

Iván Agulló, Adrián Del Río and José Navarro-Salas
On the Electric-Magnetic Duality Symmetry: Quantum Anomaly, Optical Helicity, and Particle Creation
Reprinted from: *Symmetry* **2018**, *10*, 763, doi:10.3390/sym10120763 55

István Rácz
On the Evolutionary Form of the Constraints in Electrodynamics
Reprinted from: *Symmetry* **2019**, *11*, 10, doi:10.3390/sym11010010 69

Parthasarathi Majumdar and Anarya Ray
Maxwell Electrodynamics in Terms of Physical Potentials
Reprinted from: *Symmetry* **2019**, *11*, 915, doi:10.3390/sym11070915 77

Joan Bernabeu and Jose Navarro-Salas
A Non-Local Action for Electrodynamics:Duality Symmetry and the Aharonov-Bohm Effect, Revisited
Reprinted from: *Symmetry* **2019**, *11*, 1191, doi:10.3390/sym11101191 89

Juan C. Bravo and Manuel V. Castilla
Geometric Objects: A Quality Index to Electromagnetic Energy Transfer Performance in Sustainable Smart Buildings
Reprinted from: *Symmetry* **2018**, *10*, 676, doi:10.3390/sym10120676 103

Yanping Liao, Congcong He and Qiang Guo
Denoising of Magnetocardiography Based on Improved Variational Mode Decomposition and Interval Thresholding Method
Reprinted from: *Symmetry* **2018**, *10*, 269, doi:10.3390/sym10070269 121

Abdul Raouf Al Dairy, Lina A. Al-Hmoud and Heba A. Khatatbeh
Magnetic and Structural Properties of Barium Hexaferrite Nanoparticles Doped with Titanium
Reprinted from: *Symmetry* **2019**, *11*, 732, doi:10.3390/sym11060732 . . . . **135**

Zhaoyu Guo, Danfeng Zhou, Qiang Chen, Peichang Yu and Jie Li
Design and Analysis of a Plate Type Electrodynamic Suspension Structure for Ground High Speed Systems
Reprinted from: *Symmetry* **2019**, *11*, 1117, doi:10.3390/sym11091117 . . . . **147**

Zoltán Szabó, Pavel Fiala, Jiří Zukal, Jamila Dědková and Přemysl Dohnal
Optimal Structural Design of a Magnetic Circuit for Vibration Harvesters Applicable in MEMS
Reprinted from: *Symmetry* **2020**, *12*, 110, doi:10.3390/sym12010110 . . . . **163**

About the Editors

Albert Ferrando, Full Professor, was born in València, Spain, in 1963. He received the Licenciado en Física, and M.S. and Ph.D. degrees in Theoretical Physics from the Universitat de València (UV), Burjassot, Spain, in 1985, 1986, and 1991, respectively. In 1996, he joined the Departament d'Òptica, UV, as Assistant Professor, became an Associate Professor in 2001, and Full Professor in 2011. He has developed his research in the areas of theoretical particle and condensed matter physics, optics, and microwave theory. His more recent research interests lie mainly in the electromagnetic propagation in optical waveguides, fibers, and photonic devices. The basic research interests include nonlinear optical effects in new photonic materials, quantum and mean-field effects in ultra-cold atoms, and mathematical tools for singular optics and topological photonics. His applied research includes the development of nonlinear active and passive photonic devices and the implementation of new strategies for the control of phase singularities.

Miguel Ángel García-March, Investigador Distinguido Beatriz Galindo, was born in Reus, Spain, in 1976. He received his degree in Economics from the University of Valencia in 1998. He completed his degree in Engineering as well as M.S. and Ph.D. degrees, both in Mathematical Physics, from Polytechnic University of Valencia in 2003, 2005, and 2008, respectively. He received a MEC/Fulbright two-year grant in 2009, which he held at the Colorado School of Mines. He successively held postdoctoral positions at the University College Cork (Ireland) and University of Barcelona (Spain). Between 2014 and 2019, he was Research Fellow in the Group of Maciej Lewenstein in ICFO—The Institute of Photonic Sciences. He joined the Department of Applied Mathematics of the Polytechnic University of Valencia in 2019. He has developed research in nonlinear, singular, and quantum optics; ultracold atoms; complex classical systems; and open quantum systems.

Preface to "Symmetry in Electromagnetism"

In this Special Issue, we focus on the modern view of electromagnetism, which represents both an arena for academic advance and exciting applications. This Special Issue will include contributions on electromagnetic phenomena in which symmetry plays a significant role, from a more theoretical to more applied perspectives.

Albert Ferrando, Miguel Ángel García-March
Editors

Editorial

Symmetry in Electromagnetism

Albert Ferrando [1] and Miguel Ángel García-March [2,*]

[1] Departament d'Òptica, Interdisciplinary Modeling Group, InterTech, Universitat de València, 46100 Burjassot (València), Spain; albert.ferrando@uv.es

[2] Instituto Universitario de Matemática Pura y Aplicada, Universitat Politècnica de València, E-46022 València, Spain

* Correspondence: garciamarch@mat.upv.es

Received: 21 April 2020; Accepted: 22 April 2020; Published: 26 April 2020

Electromagnetism plays an essential role, both in basic and applied physics research. The discovery of electromagnetism as the unifying theory for electricity and magnetism represented a cornerstone in modern physics. From the very beginning, symmetry was crucial to the concept of unification: Electromagnetism was soon formulated as a gauge theory, in which a local phase symmetry explained its mathematical formulation. This early connection between symmetry and electromagnetism shows that a symmetry-based approach to many electromagnetic phenomena is recurrent, even today.

Moreover, many crucial technological advances associated with electromagnetism have shaped modern civilization. The control of electromagnetic radiation in nearly all its spectra and scales is still a matter of deep interest. With the advances in material science, even at the nanoscale, the manipulation of matter–radiation interactions has reached unprecedented levels of sophistication. New generations of composite materials present effective electromagnetic properties that permit the molding of electromagnetic radiation in ways that were unconceivable just a few years ago. This is a fertile field for applications and for basic understanding in which symmetry, as in the past, bridges apparently unrelated phenomena, from condensed matter to high-energy physics.

Symmetry is the key tool in the contributions included in this Special Issue. In the context of electromagnetism, the approaches based on symmetry very often lead to diverse treatments of orbital angular momentum or pseudomomentum (as defined in e.g., [1,2]). In this direction, the most sophisticated modern approaches discuss the vectorial case, and in [3], the authors include spin-orbit coupling in nonparaxial fields, and perform a complete an analytical study of the case. The study of electromagnetic knots is also connected to orbital angular momentum, which are a consequence of applying topology concepts to Maxwell equations; in [4] the authors apply symmetry transformations to a particular electromagnetic knot, the hopfion field, to obtain a new set of knotted solutions with the properties of null. Very related to the properties of orbital angular momentum (see [1]) are periodic structures, which play a prominent role in many electromagnetic systems, e.g., microwave and antenna devices. In [5] a method to obtain the relevant transmission, reflection or absorption characteristics of a device obtained from the dispersion diagram are introduced, using general purpose electromagnetic simulation software. Digging deeply into the theory, in [6] the authors present a thorough study of quantum anomalies, which occur when a symmetry of a classical field theory is not also a symmetry of its quantum version. This is discussed in the context of a new example for quantum electromagnetic fields propagating in the presence of gravity, and applications for information extraction ARE foreseen. In this direction, constraint equations in Maxwell theory are discussed in [7]. Interestingly, this work is set in the context of an analogy with constraints of general relativity. A very deep analysis of a fully relativistically covariant and gauge-invariant formulation of classical Maxwell electrodynamics is included in [8], where the authors show the relationship of the symmetry of the inhomogeneous equations obtained and that of Minkowski spacetime. Of a great theoretical interest is also the work presented in [9], where the authors elaborate and improve the previous proposal of a nonlocal action functional for electrodynamics depending on the electric and magnetic fields, instead of potentials. They then use this formalism to confront the electric–magnetic duality symmetry of the electromagnetic field and the Aharonov–Bohm effect, two subtle aspects of electrodynamics.

Also, this book includes many applications, such as in sustainable smart buildings [10], or in magnetocardiography, where in [11] the authors present an improved variational mode decomposition model used to decompose the nonstationary signal. The magnetic properties of barium hexaferrite doped with titanium were studied in [12], where the authors propose that they could be used in the recording equipment and permanent magnets. The application to high speed systems is very appealing, such as those related to the Hyperloop concept; in particular in [13], the design and analysis of a plate-type electrodynamic suspension structure for the ground high-speed system is introduced. Finally, a report on the results of research into a vibration-powered milli-or micro-generator is given in [14], where the generators harvest mechanical energy at an optimum level, utilizing the vibration of its mechanical system; here, the authors compare some of the published microgenerator concepts and design versions by using effective power density, among other parameters, and they also provide complementary comments on the applied harvesting techniques.

This book includes papers focusing on detailed and deep theoretical studies to cutting edge applications, with many of the papers includED ALREADY harvesting many citations. The fruitful study of symmetry in electromagnetism continues to offer many encouraging surpriseS, both at a basic and an applied level.

Author Contributions: Both authors contributed equally to this work. All authors have read and agreed to the published version of the manuscript.

Funding: MAGM acknowledges funding from the Spanish Ministry of Education and Vocational Training (MEFP) through the Beatriz Galindo program 2018 (BEAGAL18/00203). A.F. acknowledges funding by the Spanish MINECO grant number TEC2017-86102-C2-1) and Generalitat Valenciana (Prometeo/2018/098).

Conflicts of Interest: The authors declare no conflict of interest.

References

1. Ferrando. Discrete-symmetry vortices as angular Bloch modes. *Phys. Rev. E* **2005**, *72*, 036612. [CrossRef] [PubMed]
2. García-March, M.A.; Ferrando, A.; Zacarés, M.; Vijande, J.; Carr, L.D. Angular pseudomomentum theory for the generalized nonlinear Schrödinger equation in discrete rotational symmetry media. *Phys. D Nonlinear Phenom.* **2009**, *238*, 1432–1438. [CrossRef]
3. Arrayás, M.; Trueba, J.L. Spin-Orbital Momentum Decomposition and Helicity Exchange in a Set of Non-Null Knotted Electromagnetic Fields. *Symmetry* **2018**, *10*, 88. [CrossRef]
4. Arrayás, M.; Rañada, A.F.; Tiemblo, A.; Trueba, J.L. Null Electromagnetic Fields from Dilatation and Rotation Transformations of the Hopfion. *Symmetry* **2019**, *11*, 1105. [CrossRef]
5. Mesa, F.; Rodríguez-Berral, R.; Medina, F. On the Computation of the Dispersion Diagram of Symmetric One-Dimensionally Periodic Structures. *Symmetry* **2018**, *10*, 307. [CrossRef]
6. Agulló, I.; del Río, A.; Navarro-Salas, J. On the Electric-Magnetic Duality Symmetry: Quantum Anomaly, Optical Helicity, and Particle Creation. *Symmetry* **2018**, *10*, 763. [CrossRef]
7. Rácz, I. On the Evolutionary Form of the Constraints in Electrodynamics. *Symmetry* **2019**, *11*, 10. [CrossRef]
8. Majumdar, P.; Ray, A. Maxwell Electrodynamics in Terms of Physical Potentials. *Symmetry* **2019**, *11*, 915. [CrossRef]
9. Bernabeu, J.; Navarro-Salas, J. A Non-Local Action for Electrodynamics: Duality Symmetry and the Aharonov-Bohm Effect, Revisited. *Symmetry* **2019**, *11*, 1191. [CrossRef]
10. Bravo, J.C.; Castilla, M.V. Geometric Objects: A Quality Index to Electromagnetic Energy Transfer Performance in Sustainable Smart Buildings. *Symmetry* **2018**, *10*, 676. [CrossRef]
11. Liao, Y.; He, C.; Guo, Q. Denoising of Magnetocardiography Based on Improved Variational Mode Decomposition and Interval Thresholding Method. *Symmetry* **2018**, *10*, 269. [CrossRef]
12. Al Dairy, A.R.; Al-Hmoud, L.A.; Khatatbeh, H.A. Magnetic and Structural Properties of Barium Hexaferrite Nanoparticles Doped with Titanium. *Symmetry* **2019**, *11*, 732. [CrossRef]

13. Guo, Z.; Zhou, D.; Chen, Q.; Yu, P.; Li, J. Design and Analysis of a Plate Type Electrodynamic Suspension Structure for Ground High Speed Systems. *Symmetry* **2019**, *11*, 1117. [CrossRef]
14. Szabó, Z.; Fiala, P.; Zukal, J.; Dědková, J.; Dohnal, P. Optimal Structural Design of a Magnetic Circuit for Vibration Harvesters Applicable in MEMS. *Symmetry* **2020**, *12*, 110. [CrossRef]

© 2020 by the authors. Licensee MDPI, Basel, Switzerland. This article is an open access article distributed under the terms and conditions of the Creative Commons Attribution (CC BY) license (http://creativecommons.org/licenses/by/4.0/).

Article

Spin-Orbital Momentum Decomposition and Helicity Exchange in a Set of Non-Null Knotted Electromagnetic Fields

Manuel Arrayás [†,*] and José L. Trueba [†]

Área de Electromagnetismo, Universidad Rey Juan Carlos, Calle Tulipán s/n, 28933 Móstoles (Madrid), Spain; joseluis.trueba@urjc.es
* Correspondence: manuel.arrayas@urjc.es
† These authors contributed equally to this work.

Received: 9 March 2018; Accepted: 27 March 2018; Published: 30 March 2018

Abstract: We calculate analytically the spin-orbital decomposition of the angular momentum using completely nonparaxial fields that have a certain degree of linkage of electric and magnetic lines. The split of the angular momentum into spin-orbital components is worked out for non-null knotted electromagnetic fields. The relation between magnetic and electric helicities and spin-orbital decomposition of the angular momentum is considered. We demonstrate that even if the total angular momentum and the values of the spin and orbital momentum are the same, the behavior of the local angular momentum density is rather different. By taking cases with constant and non-constant electric and magnetic helicities, we show that the total angular momentum density presents different characteristics during time evolution.

Keywords: electromagnetic knots; helicity; spin-orbital momentum

1. Introduction

There has been recently some interest in the orbital-spin decomposition of the angular momentum carried by light. The total angular momentum can be decomposed into orbital and spin angular momenta for paraxial light, but for nonparaxial fields, that splitting is more controversial because their quantized forms do not satisfy the commutation relations [1,2]. For a review and references, see for example [3,4].

In this work, we provide an exact calculation of the orbital-spin decomposition of the angular momentum in a completely nonparaxial field. We compute the orbital-spin contributions to the total angular momentum analytically for a knotted class of fields [5]. These fields have nontrivial electromagnetic helicity [6,7]. We show that the existence of electromagnetic fields in a vacuum with the same constant angular momentum and orbital-spin decomposition, but different electric and magnetic helicities is possible. We find cases where the helicities are constant during the field evolution and cases where they change in time, evolving through a phenomenon of exchanging magnetic and electric components [8]. The angular momentum density presents different time evolution in each case.

The orbital-spin decomposition and its observability has been discussed in the context of the dual symmetry of Maxwell equations in a vacuum [9]. In this paper, we first make a brief review of the concept of electromagnetic duality. That duality, termed "electromagnetic democracy" [10], has been central in the work of knotted field configurations [5,11–26]. Related field configurations have also appeared in plasma physics [27–30], optics [31–35], classical field theory [36], quantum physics [37,38], various states of matter [39–43] and twistors [44,45].

We will make use of the helicity basis [7] in order to write the magnetic and electric spin of the field in that basis, which simplifies the calculations, as well as the magnetic and electric helicities' components. On that basis, we will get some general results, such as the difference between the magnetic and electric spin components in the Coulomb gauge is null. This conclusion coincides with the results found, for example, in [46] using a different approach. We proceed by giving the explicit calculation of the decomposition of the angular momentum into spin and orbital components for a whole class of fields, the non-null toroidal class [5,25]. We will show that the angular decomposition remains constant in time, while the helicities may or may not change. We provide an example of each case and plot the time evolution of the total angular momentum density. In the final section, we summarize the main results.

2. Duality and Helicity in Maxwell Theory in a Vacuum

In this section, we will review the definition of magnetic and electric helicities. These definitions are possible because of the dual property of electromagnetism in a vacuum. We will also describe a vector density, which can be identified with the spin density using the helicity four-current zeroth component.

Electromagnetism in a vacuum can be described in terms of two real vector fields, **E** and **B**, called the electric and magnetic fields, respectively. Using the SI units, these fields satisfy Maxwell equations in a vacuum,

$$\nabla \cdot \mathbf{B} = 0, \quad \nabla \times \mathbf{E} + \partial_t \mathbf{B} = 0, \tag{1}$$

$$\nabla \cdot \mathbf{E} = 0, \quad \nabla \times \mathbf{B} - \frac{1}{c^2} \partial_t \mathbf{E} = 0. \tag{2}$$

Using the four-vector electromagnetic potential:

$$A^\mu = \left(\frac{V}{c}, \mathbf{A}\right), \tag{3}$$

where V and $\mathbf{A}$ are the scalar and vector potential, respectively, the electromagnetic field tensor is:

$$F_{\mu\nu} = \partial_\mu A_\nu - \partial_\nu A_\mu. \tag{4}$$

From Equation (4), the electric and magnetic field components are:

$$\mathbf{E}_i = c\, F^{i0}, \quad \mathbf{B}_i = -\frac{1}{2} \epsilon_{ijk} F^{jk}, \tag{5}$$

or, in three-dimensional quantities,

$$\mathbf{E} = -\nabla V - \frac{\partial \mathbf{A}}{\partial t}, \quad \mathbf{B} = \nabla \times \mathbf{A}. \tag{6}$$

Since Equation (1) is just identities in terms of the four-vector electromagnetic potential Equation (3), by using (6), the dynamics of electromagnetism is given by Equation (2), which can be written as:

$$\partial_\mu F^{\mu\nu} = 0. \tag{7}$$

Partly based on the duality property of Maxwell equations in a vacuum [47], there is the idea of "electromagnetic democracy" [9,10]. The equations are invariant under the map $(\mathbf{E}, c\mathbf{B}) \mapsto (c\mathbf{B}, -\mathbf{E})$. Electromagnetic democracy means that, in a vacuum, it is possible to define another four-potential:

$$C^\mu = (c\, V', \mathbf{C}), \tag{8}$$

so that the dual of the electromagnetic tensor $F_{\mu\nu}$ in Equation (4), defined as:

$$^*F_{\mu\nu} = \frac{1}{2}\varepsilon_{\mu\nu\alpha\beta}F^{\alpha\beta},\qquad(9)$$

satisfies:

$$^*F_{\mu\nu} = -\frac{1}{c}\left(\partial_\mu C_\nu - \partial_\nu C_\mu\right),\qquad(10)$$

or, in terms of three-dimensional fields,

$$\mathbf{E} = \nabla \times \mathbf{C},\quad \mathbf{B} = \nabla V' + \frac{1}{c^2}\frac{\partial \mathbf{C}}{\partial t}.\qquad(11)$$

Equation (2) is again identities when the definitions (11) are imposed. Thus, Maxwell equations in a vacuum can be described in terms of two sets of vector potentials as in definition Equations (4) and (10), which have to satisfy the duality condition Equation (9).

In the study of topological configurations of electric and magnetic lines, an important quantity is the helicity of a vector field [48–53], which can be defined for every divergenceless three-dimensional vector field. Magnetic helicity is related to the linkage of magnetic lines. In the case of electromagnetism in a vacuum, the magnetic helicity can be defined as the integral:

$$h_m = \frac{1}{2c\mu_0}\int d^3r\, \mathbf{A}\cdot \mathbf{B},\qquad(12)$$

where c is the speed of light in a vacuum and μ_0 is the vacuum permeability. Note that, in this equation, the magnetic helicity is taken so that it has dimensions of angular momentum in SI units. Since the electric field in a vacuum is also divergenceless, an electric helicity, related to the linking number of electric lines, can also be defined as:

$$h_e = \frac{\varepsilon_0}{2c}\int d^3r\, \mathbf{C}\cdot \mathbf{E} = \frac{1}{2c^3\mu_0}\int d^3r\, \mathbf{C}\cdot \mathbf{E},\qquad(13)$$

where $\varepsilon_0 = 1/(c^2\mu_0)$ is the vacuum electric permittivity. Electric helicity in Equation (13) also has dimensions of angular momentum. Magnetic and electric helicities in a vacuum can be studied in terms of helicity four-currents [6,7,9,17], so that the magnetic helicity density is the zeroth component of:

$$\mathcal{H}_m^\mu = -\frac{1}{2c\mu_0}A_\nu\,^*F^{\nu\mu},\qquad(14)$$

and the electric helicity is the zeroth component of:

$$\mathcal{H}_e^\mu = -\frac{1}{2c^2\mu_0}C_\nu F^{\nu\mu}.\qquad(15)$$

The divergence of $\mathcal{H}_m^\mu$ and $\mathcal{H}_e^\mu$ is related to the time conservation of both helicities,

$$\begin{aligned}\partial_\mu \mathcal{H}_m^\mu &= \frac{1}{4c\mu_0}F_{\mu\nu}\,^*F^{\mu\nu},\\ \partial_\mu \mathcal{H}_e^\mu &= -\frac{1}{4c\mu_0}\,^*F_{\mu\nu}F^{\mu\nu},\end{aligned}\qquad(16)$$

which yields:

$$\frac{dh_m}{dt} = -\frac{1}{2c\mu_0} \int (V\mathbf{B} - \mathbf{A} \times \mathbf{E}) \cdot d\mathbf{S} - \frac{1}{c\mu_0} \int d^3r\, \mathbf{E} \cdot \mathbf{B},$$
$$\frac{dh_e}{dt} = -\frac{1}{2c\mu_0} \int (V'\mathbf{E} + \mathbf{C} \times \mathbf{B}) \cdot d\mathbf{S} + \frac{1}{c\mu_0} \int d^3r\, \mathbf{E} \cdot \mathbf{B}. \quad (17)$$

In the special case that the domain of integration of Equation (17) is the whole R^3 space and the fields behave at infinity in a way such that the surface integrals in Equation (17) vanish, we get:

- If the integral of $\mathbf{E} \cdot \mathbf{B}$ is zero, both the magnetic and the electric helicities are constant during the evolution of the electromagnetic field.
- If the integral of $\mathbf{E} \cdot \mathbf{B}$ is not zero, the helicities are not constant, but they satisfy:

$$\frac{dh_m}{dt} = -\frac{dh_e}{dt}, \quad (18)$$

so there is an interchange of helicities between the magnetic and electric parts of the field [8].
- For every value of the integral of $\mathbf{E} \cdot \mathbf{B}$, the electromagnetic helicity h, defined as:

$$h = h_m + h_e = \frac{1}{2c\mu_0} \int d^3r\, \mathbf{A} \cdot \mathbf{B} + \frac{\varepsilon_0}{2c} \int d^3r\, \mathbf{C} \cdot \mathbf{E}, \quad (19)$$

is a conserved quantity.

If the domain of integration of Equation (17) is restricted to a finite volume Ω, then the flux of electromagnetic helicity through the boundary $\partial\Omega$ of the volume is given by:

$$\frac{dh}{dt} = -\frac{1}{2c\mu_0} \int_{\partial\Omega} \left[(V\mathbf{B} - \mathbf{A} \times \mathbf{E}) + (V'\mathbf{E} + \mathbf{C} \times \mathbf{B}) \right] \cdot d\mathbf{S}. \quad (20)$$

The integrand in the second term of this equation defines a vector density whose components are given by $\mathbf{S}_i = \mathcal{H}_m^i + \mathcal{H}_e^i$, so that:

$$\mathbf{S} = \frac{1}{2c^2\mu_0} \left(V\mathbf{B} - \mathbf{A} \times \mathbf{E} + V'\mathbf{E} + \mathbf{C} \times \mathbf{B} \right). \quad (21)$$

This vector density has been considered as a physically meaningful spin density for the electromagnetic field in a vacuum in some references [9,46,54–56]. In the following, we examine some questions about the relation between the magnetic and electric parts of the helicity and their corresponding magnetic and electric parts of the spin.

3. Fourier Decomposition and Helicity Basis for the Electromagnetic Field in a Vacuum

In this section, we will write the electromagnetic fields in terms of the helicity basis, which will be very useful for obtaining the results and computations presented in the following sections.

The electric and magnetic fields can be decomposed into Fourier terms,

$$\mathbf{E}(\mathbf{r},t) = \frac{1}{(2\pi)^{3/2}} \int d^3k \left(\mathbf{E}_1(\mathbf{k}) e^{-ikx} + \mathbf{E}_2(\mathbf{k}) e^{ikx} \right),$$
$$\mathbf{B}(\mathbf{r},t) = \frac{1}{(2\pi)^{3/2}} \int d^3k \left(\mathbf{B}_1(\mathbf{k}) e^{-ikx} + \mathbf{B}_2(\mathbf{k}) e^{ikx} \right), \quad (22)$$

where we have introduced the four-dimensional notation $kx = \omega t - \mathbf{k} \cdot \mathbf{r}$, with $\omega = kc$.

For the vector potentials, we need to fix a gauge. In the Coulomb gauge, the vector potentials are chosen so that $V = 0$, $\nabla \cdot \mathbf{A} = 0$, $V' = 0$, $\nabla \cdot \mathbf{C} = 0$. Then, they satisfy the relations:

$$\mathbf{B} = \nabla \times \mathbf{A} = \frac{1}{c^2} \frac{\partial \mathbf{C}}{\partial t},$$

$$\mathbf{E} = \nabla \times \mathbf{C} = -\frac{\partial \mathbf{A}}{\partial t}. \tag{23}$$

One can write for them the following Fourier decomposition,

$$\mathbf{A}(\mathbf{r}, t) = \frac{1}{(2\pi)^{3/2}} \int d^3k \left[e^{-ikx} \bar{\mathbf{a}}(\mathbf{k}) + e^{ikx} \mathbf{a}(\mathbf{k}) \right],$$

$$\mathbf{C}(\mathbf{r}, t) = \frac{c}{(2\pi)^{3/2}} \int d^3k \left[e^{-ikx} \bar{\mathbf{c}}(\mathbf{k}) + e^{ikx} \mathbf{c}(\mathbf{k}) \right], \tag{24}$$

where the factor c in $\mathbf{C}$ is taken for dimensional reasons and $\bar{\mathbf{a}}, \bar{\mathbf{c}}$ denotes the complex conjugate of $\mathbf{a}, \mathbf{c}$, respectively. Taking time derivatives and using the Coulomb gauge conditions Equation (23),

$$\mathbf{E} = -\frac{\partial \mathbf{A}}{\partial t} = \frac{1}{(2\pi)^{3/2}} \int d^3k \left[e^{-ikx} (ikc) \bar{\mathbf{a}}(\mathbf{k}) - e^{ikx} (ikc) \mathbf{a}(\mathbf{k}) \right],$$

$$\mathbf{B} = \frac{1}{c^2} \frac{\partial \mathbf{C}}{\partial t} = \frac{1}{(2\pi)^{3/2}} \int d^3k \left[-e^{-ikx} (ik) \bar{\mathbf{c}}(\mathbf{k}) + e^{ikx} (ik) \mathbf{c}(\mathbf{k}) \right]. \tag{25}$$

and by comparison with Equation (22), one can get the values for $\mathbf{a}(\mathbf{k})$ and $\mathbf{c}(\mathbf{k})$.

The helicity Fourier components appear when the vector potentials $\mathbf{A}$ and $\mathbf{C}$, in the Coulomb gauge, are written as a combination of circularly-polarized plane waves [57], as:

$$\mathbf{A}(\mathbf{r},t) = \frac{\sqrt{\hbar c \mu_0}}{(2\pi)^{3/2}} \int \frac{d^3k}{\sqrt{2k}} \left[e^{-ikx} (a_R(\mathbf{k}) \mathbf{e}_R(\mathbf{k}) + a_L(\mathbf{k}) \mathbf{e}_L(\mathbf{k})) + C.C \right],$$

$$\mathbf{C}(\mathbf{r},t) = \frac{c\sqrt{\hbar c \mu_0}}{(2\pi)^{3/2}} \int \frac{d^3k}{\sqrt{2k}} \left[i e^{-ikx} (a_R(\mathbf{k}) \mathbf{e}_R(\mathbf{k}) - a_L(\mathbf{k}) \mathbf{e}_L(\mathbf{k})) + C.C \right]. \tag{26}$$

where $\hbar$ is the Planck constant and $C.C$ means the complex conjugate. The Fourier components in the helicity basis are given by the unit vectors $\mathbf{e}_R(\mathbf{k})$, $\mathbf{e}_L(\mathbf{k})$, $\mathbf{e}_k = \mathbf{k}/k$, and the helicity components $a_R(\mathbf{k})$, $a_L(\mathbf{k})$ that, in the quantum theory, are interpreted as annihilation operators of photon states with right- and left-handed polarization, respectively. In quantum theory, $\bar{a}_R(\mathbf{k})$, $\bar{a}_L(\mathbf{k})$ are creation operators of such states.

In order to simplify the notation, most of the time, we will not write explicitly the dependence on $\mathbf{k}$ of the basis vectors and coefficients, meaning $a_L = a_L(\mathbf{k})$, $\mathbf{e}_R = \mathbf{e}_R(\mathbf{k})$, $a'_L = a_L(\mathbf{k}')$, $\mathbf{e}'_R = \mathbf{e}_R(\mathbf{k}')$.

The unit vectors in the helicity basis are taken to satisfy:

$$\begin{aligned} &\bar{\mathbf{e}}_R = \mathbf{e}_L, \ \mathbf{e}_R(-\mathbf{k}) = -\mathbf{e}_L(\mathbf{k}), \ \mathbf{e}_L(-\mathbf{k}) = -\mathbf{e}_R(\mathbf{k}), \\ &\mathbf{e}_k \cdot \mathbf{e}_R = \mathbf{e}_k \cdot \mathbf{e}_L = 0, \ \mathbf{e}_R \cdot \mathbf{e}_R = \mathbf{e}_L \cdot \mathbf{e}_L = 0, \ \mathbf{e}_R \cdot \mathbf{e}_L = 1, \\ &\mathbf{e}_k \times \mathbf{e}_k = \mathbf{e}_R \times \mathbf{e}_R = \mathbf{e}_L \times \mathbf{e}_L = 0, \\ &\mathbf{e}_k \times \mathbf{e}_R = -i\mathbf{e}_R, \ \mathbf{e}_k \times \mathbf{e}_L = i\mathbf{e}_L, \ \mathbf{e}_R \times \mathbf{e}_L = -i\mathbf{e}_k, \end{aligned} \tag{27}$$

The relation between the helicity basis and the planar Fourier basis can be obtained by comparing Equations (24) and (26). Consequently, the electric and magnetic fields of an electromagnetic field in a vacuum, and the vector potentials in the Coulomb gauge can be expressed in this basis as:

$$\mathbf{E}(\mathbf{r},t) = \frac{ic\sqrt{\hbar c \mu_0}}{(2\pi)^{3/2}} \int d^3k \sqrt{\frac{k}{2}} \left[e^{-ikx}(a_R \mathbf{e}_R + a_L \mathbf{e}_L) - e^{ikx}(\bar{a}_R \mathbf{e}_L + \bar{a}_L \mathbf{e}_R) \right]$$

$$\mathbf{B}(\mathbf{r},t) = \frac{\sqrt{\hbar c \mu_0}}{(2\pi)^{3/2}} \int d^3k \sqrt{\frac{k}{2}} \left[e^{-ikx}(a_R \mathbf{e}_R - a_L \mathbf{e}_L) + e^{ikx}(\bar{a}_R \mathbf{e}_L - \bar{a}_L \mathbf{e}_R) \right]$$

$$\mathbf{A}(\mathbf{r},t) = \frac{\sqrt{\hbar c \mu_0}}{(2\pi)^{3/2}} \int d^3k \frac{1}{\sqrt{2k}} \left[e^{-ikx}(a_R \mathbf{e}_R + a_L \mathbf{e}_L) + e^{ikx}(\bar{a}_R \mathbf{e}_L + \bar{a}_L \mathbf{e}_R) \right]$$

$$\mathbf{C}(\mathbf{r},t) = \frac{ic\sqrt{\hbar c \mu_0}}{(2\pi)^{3/2}} \int d^3k \frac{1}{\sqrt{2k}} \left[e^{-ikx}(a_R \mathbf{e}_R - a_L \mathbf{e}_L) - e^{ikx}(\bar{a}_R \mathbf{e}_L - \bar{a}_L \mathbf{e}_R) \right] \tag{28}$$

where the unit vectors satisfy the relations Equation (27).

It is interesting to point the fact that in the helicity basis, we get for the magnetic vector potential the relation:

$$\mathbf{A}(k) = -\frac{\mathbf{k} \times \mathbf{k} \times \mathbf{A}(k)}{\mathbf{k} \cdot \mathbf{k}}, \tag{29}$$

where:

$$\mathbf{A}(k) = e^{-ikx}(a_R \mathbf{e}_R + a_L \mathbf{e}_L) + e^{ikx}(\bar{a}_R \mathbf{e}_L + \bar{a}_L \mathbf{e}_R), \tag{30}$$

taken from Equation (28). In reference Equation [58], the nonlocality of electromagnetic quantities is discussed, and the transverse part of Fourier components of the vector potential is introduced as:

$$\mathbf{A}^{\perp}(k) = -\frac{\mathbf{k} \times \mathbf{k} \times \mathbf{A}(k)}{\mathbf{k} \cdot \mathbf{k}}. \tag{31}$$

We can see explicitly now from Equations (29) and (31) that in the Coulomb gauge in the helicity basis:

$$\mathbf{A}(k) = \mathbf{A}^{\perp}(k).$$

4. Magnetic and Electric Helicities in the Helicity Basis

In the previous section, we have introduced the helicity basis and expressed the fields in that basis. In this section, we will express the electric and magnetic helicities in the same basis [7].

If we use the expressions (28), the magnetic helicity can be written as:

$$\begin{aligned} h_m &= \frac{1}{2c\mu_0} \int d^3r\, \mathbf{A} \cdot \mathbf{B} = \frac{\hbar}{4} \int d^3k \int d^3k' \int \frac{d^3r}{(2\pi)^3} \sqrt{\frac{k'}{k}} \\ &\quad \left[e^{-i\omega t} e^{i\omega' t} e^{i(\mathbf{k}-\mathbf{k}')\cdot\mathbf{r}} (a_R \mathbf{e}_R + a_L \mathbf{e}_L) \cdot (\bar{a}'_R \mathbf{e}'_L - \bar{a}'_L \mathbf{e}'_R) \right. \\ &\quad + e^{i\omega t} e^{-i\omega' t} e^{-i(\mathbf{k}-\mathbf{k}')\cdot\mathbf{r}} (\bar{a}_R \mathbf{e}_L + \bar{a}_L \mathbf{e}_R) \cdot (a'_R \mathbf{e}'_R - a'_L \mathbf{e}'_L) \\ &\quad + e^{-i\omega t} e^{-i\omega' t} e^{i(\mathbf{k}+\mathbf{k}')\cdot\mathbf{r}} (a_R \mathbf{e}_R + a_L \mathbf{e}_L) \cdot (a'_R \mathbf{e}'_R - a'_L \mathbf{e}'_L) \\ &\quad \left. + e^{i\omega t} e^{i\omega' t} e^{-i(\mathbf{k}+\mathbf{k}')\cdot\mathbf{r}} (\bar{a}_R \mathbf{e}_L + \bar{a}_L \mathbf{e}_R) \cdot (\bar{a}'_R \mathbf{e}'_L - \bar{a}'_L \mathbf{e}'_R) \right]. \end{aligned} \tag{32}$$

Taking into account the following property of the Dirac-delta function,

$$\int d^3k' \int \frac{d^3r}{(2\pi)^3} e^{-i(\mathbf{k}-\mathbf{k}')\cdot\mathbf{r}} \left(\mathbf{f}(\mathbf{k}) \cdot \mathbf{g}(\mathbf{k}') \right) = \mathbf{f}(\mathbf{k}) \cdot \mathbf{g}(\mathbf{k}), \tag{33}$$

and using the relations (27) yields:

$$\begin{aligned} h_m &= \frac{\hbar}{2} \int d^3k \, (\bar{a}_R(\mathbf{k}) a_R(\mathbf{k}) - \bar{a}_L(\mathbf{k}) a_L(\mathbf{k})) \\ &+ \frac{\hbar}{4} \int d^3k \, e^{-2i\omega t} \, (-a_R(\mathbf{k}) a_R(-\mathbf{k}) + a_L(\mathbf{k}) a_L(-\mathbf{k})) \\ &+ \frac{\hbar}{4} \int d^3k \, e^{2i\omega t} \, (-\bar{a}_R(\mathbf{k}) \bar{a}_R(-\mathbf{k}) + \bar{a}_L(\mathbf{k}) \bar{a}_L(-\mathbf{k})) \,. \end{aligned} \qquad (34)$$

We observe that the magnetic helicity has two contributions: the first term in Equation (34) is independent of time, and the rest of the terms constitute the time-dependent part of the magnetic helicity.

We repeat the same procedure for the electric helicity. The electric helicity can be written as:

$$\begin{aligned} h_e &= \frac{1}{2c^3 \mu_0} \int d^3r \, \mathbf{C} \cdot \mathbf{E} = \frac{\hbar}{4} \int d^3k \int d^3k' \int \frac{d^3r}{(2\pi)^3} \sqrt{\frac{k'}{k}} \\ &\quad \Big[e^{-i\omega t} e^{i\omega' t} e^{i(\mathbf{k}-\mathbf{k}')\cdot \mathbf{r}} \, (a_R \mathbf{e}_R - a_L \mathbf{e}_L) \cdot (\bar{a}'_R \mathbf{e}'_L + \bar{a}'_L \mathbf{e}'_R) \\ &\quad + e^{i\omega t} e^{-i\omega' t} e^{-i(\mathbf{k}-\mathbf{k}')\cdot \mathbf{r}} \, (\bar{a}_R \mathbf{e}_L - \bar{a}_L \mathbf{e}_R) \cdot (a'_R \mathbf{e}'_R + a'_L \mathbf{e}'_L) \\ &\quad - e^{-i\omega t} e^{-i\omega' t} e^{i(\mathbf{k}+\mathbf{k}')\cdot \mathbf{r}} \, (a_R \mathbf{e}_R - a_L \mathbf{e}_L) \cdot (a'_R \mathbf{e}'_R + a'_L \mathbf{e}'_L) \\ &\quad - e^{i\omega t} e^{i\omega' t} e^{-i(\mathbf{k}+\mathbf{k}')\cdot \mathbf{r}} \, (\bar{a}_R \mathbf{e}_L - \bar{a}_L \mathbf{e}_R) \cdot (\bar{a}'_R \mathbf{e}'_L + \bar{a}'_L \mathbf{e}'_R) \Big] \,, \end{aligned} \qquad (35)$$

and again using Equations (33) and ((27), we get,

$$\begin{aligned} h_e &= \frac{\hbar}{2} \int d^3k \, (\bar{a}_R(\mathbf{k}) a_R(\mathbf{k}) - \bar{a}_L(\mathbf{k}) a_L(\mathbf{k})) \\ &- \frac{\hbar}{4} \int d^3k \, e^{-2i\omega t} \, (-a_R(\mathbf{k}) a_R(-\mathbf{k}) + a_L(\mathbf{k}) a_L(-\mathbf{k})) \\ &- \frac{\hbar}{4} \int d^3k \, e^{2i\omega t} \, (-\bar{a}_R(\mathbf{k}) \bar{a}_R(-\mathbf{k}) + \bar{a}_L(\mathbf{k}) \bar{a}_L(-\mathbf{k})) \,. \end{aligned} \qquad (36)$$

The electromagnetic helicity h in a vacuum is the sum of the magnetic and electric helicities. From Equations (34) and (36),

$$h = h_m + h_e = \hbar \int d^3k \, (\bar{a}_R(\mathbf{k}) a_R(\mathbf{k}) - \bar{a}_L(\mathbf{k}) a_L(\mathbf{k})) \,. \qquad (37)$$

In quantum electrodynamics, the integral in the right-hand side of Equation (37) is interpreted as the helicity operator, which subtracts the number of left-handed photons from the number of right-handed photons. From the usual expressions:

$$\begin{aligned} N_R &= \int d^3k \, \bar{a}_R(\mathbf{k}) a_R(\mathbf{k}), \\ N_L &= \int d^3k \, \bar{a}_L(\mathbf{k}) a_L(\mathbf{k}), \end{aligned} \qquad (38)$$

we can write (37) as:

$$h = \hbar \, (N_R - N_L) \,. \qquad (39)$$

Consequently, the electromagnetic helicity (19) is the classical limit of the difference between the numbers of right-handed and left-handed photons [6,7,15].

However, the difference between the magnetic and electric helicities depends on time in general, since:

$$\tilde{h}(t) = h_m - h_e = \frac{\hbar}{2} \int d^3k \left[e^{-2i\omega t} \left(-a_R(\mathbf{k})a_R(-\mathbf{k}) + a_L(\mathbf{k})a_L(-\mathbf{k}) \right) \right.$$
$$\left. + e^{2i\omega t} \left(-\bar{a}_R(\mathbf{k})\bar{a}_R(-\mathbf{k}) + \bar{a}_L(\mathbf{k})\bar{a}_L(-\mathbf{k}) \right) \right]. \tag{40}$$

so the electromagnetic field is allowed to exchange electric and magnetic helicity components during its evolution. For an account of this phenomenon, we refer to [8,25].

5. Magnetic and Electric Spin in the Helicity Basis

Now in this section, we are going to express the magnetic and electric spins components of the total angular momentum in the helicity basis.

Let us consider the spin vector defined by Equation (21). It can be written as:

$$\mathbf{s} = \mathbf{s}_m + \mathbf{s}_e, \tag{41}$$

where the magnetic part of the spin is defined from the flux of magnetic helicity,

$$\mathbf{s}_m = \frac{1}{2c^2\mu_0} \int d^3r \left(V \mathbf{B} - \mathbf{A} \times \mathbf{E} \right), \tag{42}$$

and the electric spin comes from the flux of the electric helicity,

$$\mathbf{s}_e = \frac{1}{2c^2\mu_0} \int d^3r \left(V' \mathbf{E} + \mathbf{C} \times \mathbf{B} \right). \tag{43}$$

Note that the electric spin in Equation (43) can be defined only for the case of electromagnetism in a vacuum, in the same way as the electric helicity is defined only in a vacuum.

Using the helicity basis of the previous sections, which was calculated in the Coulomb gauge, the magnetic spin can be written as:

$$\mathbf{s}_m = \frac{1}{2c^2\mu_0} \int d^3r\, \mathbf{E} \times \mathbf{A} = \frac{\hbar}{4} \int d^3k \int d^3k' \int \frac{d^3r}{(2\pi)^3} \sqrt{\frac{k'}{k}}$$
$$\left[ie^{-i\omega t}e^{i\omega' t}e^{i(\mathbf{k}-\mathbf{k}')\cdot\mathbf{r}} \left(a_R\mathbf{e}_R + a_L\mathbf{e}_L \right) \times \left(\bar{a}'_R\mathbf{e}'_L + \bar{a}'_L\mathbf{e}'_R \right) \right.$$
$$- ie^{i\omega t}e^{-i\omega' t}e^{-i(\mathbf{k}-\mathbf{k}')\cdot\mathbf{r}} \left(\bar{a}_R\mathbf{e}_L + \bar{a}_L\mathbf{e}_R \right) \times \left(a'_R\mathbf{e}'_R + a'_L\mathbf{e}'_L \right)$$
$$- ie^{-i\omega t}e^{-i\omega' t}e^{i(\mathbf{k}+\mathbf{k}')\cdot\mathbf{r}} \left(a_R\mathbf{e}_R + a_L\mathbf{e}_L \right) \times \left(a'_R\mathbf{e}'_R + a'_L\mathbf{e}'_L \right)$$
$$\left. + ie^{i\omega t}e^{i\omega' t}e^{-i(\mathbf{k}+\mathbf{k}')\cdot\mathbf{r}} \left(\bar{a}_R\mathbf{e}_L + \bar{a}_L\mathbf{e}_R \right) \times \left(\bar{a}'_R\mathbf{e}'_L + \bar{a}'_L\mathbf{e}'_R \right) \right], \tag{44}$$

and after the same manipulations as in the previous section, using Equations (33) and (27), it turns out:

$$\mathbf{s}_m = \frac{\hbar}{2} \int d^3k \left(\bar{a}_R(\mathbf{k})a_R(\mathbf{k}) - \bar{a}_L(\mathbf{k})a_L(\mathbf{k}) \right) \mathbf{e}_k$$
$$+ \frac{\hbar}{4} \int d^3k\, e^{-2i\omega t} \left(a_R(\mathbf{k})a_R(-\mathbf{k}) - a_L(\mathbf{k})a_L(-\mathbf{k}) \right) \mathbf{e}_k$$
$$+ \frac{\hbar}{4} \int d^3k\, e^{2i\omega t} \left(\bar{a}_R(\mathbf{k})\bar{a}_R(-\mathbf{k}) - \bar{a}_L(\mathbf{k})\bar{a}_L(-\mathbf{k}) \right) \mathbf{e}_k. \tag{45}$$

As in the case of magnetic helicity Equation (34), the magnetic spin has two contributions: the first term in Equation (45) is independent of time, while the rest of the terms are, in principle, time-dependent.

In a similar way, the electric spin in the helicity basis is:

$$\begin{aligned}
\mathbf{s}_e &= \frac{1}{2c^2\mu_0}\int d^3r\,\mathbf{C}\times\mathbf{B} = \frac{\hbar}{4}\int d^3k\int d^3k'\int \frac{d^3r}{(2\pi)^3}\sqrt{\frac{k'}{k}}\\
&\quad \Big[ie^{-i\omega t}e^{i\omega' t}e^{i(\mathbf{k}-\mathbf{k}')\cdot\mathbf{r}}\,(a_R\mathbf{e}_R - a_L\mathbf{e}_L)\times(\bar{a}'_R\mathbf{e}'_L - \bar{a}'_L\mathbf{e}'_R)\\
&\quad - ie^{i\omega t}e^{-i\omega' t}e^{-i(\mathbf{k}-\mathbf{k}')\cdot\mathbf{r}}\,(\bar{a}_R\mathbf{e}_L - \bar{a}_L\mathbf{e}_R)\times(a'_R\mathbf{e}'_R - a'_L\mathbf{e}'_L)\\
&\quad + ie^{-i\omega t}e^{-i\omega' t}e^{i(\mathbf{k}+\mathbf{k}')\cdot\mathbf{r}}\,(a_R\mathbf{e}_R - a_L\mathbf{e}_L)\times(a'_R\mathbf{e}'_R - a'_L\mathbf{e}'_L)\\
&\quad - ie^{i\omega t}e^{i\omega' t}e^{-i(\mathbf{k}+\mathbf{k}')\cdot\mathbf{r}}\,(\bar{a}_R\mathbf{e}_L - \bar{a}_L\mathbf{e}_R)\times(\bar{a}'_R\mathbf{e}'_L - \bar{a}'_L\mathbf{e}'_R)\Big],
\end{aligned} \qquad (46)$$

that after integrating in $\mathbf{k}'$ gives:

$$\begin{aligned}
\mathbf{s}_e &= \frac{\hbar}{2}\int d^3k\,(\bar{a}_R(\mathbf{k})a_R(\mathbf{k}) - \bar{a}_L(\mathbf{k})a_L(\mathbf{k}))\,\mathbf{e}_k\\
&\quad + \frac{\hbar}{4}\int d^3k\,e^{-2i\omega t}\,(-a_R(\mathbf{k})a_R(-\mathbf{k}) + a_L(\mathbf{k})a_L(-\mathbf{k}))\,\mathbf{e}_k\\
&\quad + \frac{\hbar}{4}\int d^3k\,e^{2i\omega t}\,(-\bar{a}_R(\mathbf{k})\bar{a}_R(-\mathbf{k}) + \bar{a}_L(\mathbf{k})\bar{a}_L(-\mathbf{k}))\,\mathbf{e}_k.
\end{aligned} \qquad (47)$$

Finally, the spin of the electromagnetic field in a vacuum is, according to Equation (41),

$$\mathbf{s} = \mathbf{s}_m + \mathbf{s}_e = \hbar\int d^3k\,(\bar{a}_R(\mathbf{k})a_R(\mathbf{k}) - \bar{a}_L(\mathbf{k})a_L(\mathbf{k}))\,\mathbf{e}_k, \qquad (48)$$

an expression that is equivalent to the well-known result in quantum electrodynamics [57].

We can compute, as we did for the helicity, the difference between the magnetic and electric parts of the spin,

$$\tilde{\mathbf{s}}(t) = \mathbf{s}_m - \mathbf{s}_e = \tfrac{\hbar}{2}\int d^3k\,\big[e^{-2i\omega t}(a_R(\mathbf{k})a_R(-\mathbf{k}) - a_L(\mathbf{k})a_L(-\mathbf{k})) + e^{2i\omega t}(\bar{a}_R(\mathbf{k})\bar{a}_R(-\mathbf{k}) - \bar{a}_L(\mathbf{k})\bar{a}_L(-\mathbf{k}))\big]\,\mathbf{e}_k. \qquad (49)$$

Note the similarity in the integrands of the difference between helicities Equation (40) and the difference between spins (49). Both have one term proportional to the complex quantity:

$$f(\mathbf{k}) = a_R(\mathbf{k})a_R(-\mathbf{k}) - a_L(\mathbf{k})a_L(-\mathbf{k}), \qquad (50)$$

and another term proportional to the complex conjugate of $f(\mathbf{k})$. It is obvious that $f(\mathbf{k})$ is an even function of the wave vector $\mathbf{k}$. This means, in particular, that the integral Equation (49) is identically zero, so the spin difference satisfies:

$$\tilde{\mathbf{s}}(t) = 0. \qquad (51)$$

Thus, we arrive at the following result for any electromagnetic field in a vacuum,

$$\mathbf{s}_m = \mathbf{s}_e = \frac{1}{2}\mathbf{s} = \frac{\hbar}{2}\int d^3k\,(\bar{a}_R(\mathbf{k})a_R(\mathbf{k}) - \bar{a}_L(\mathbf{k})a_L(\mathbf{k}))\,\mathbf{e}_k. \qquad (52)$$

This conclusion coincides with the results found in [46].

Therefore, while the magnetic and electric spins are equal in electromagnetism in a vacuum, in general, this fact does not apply to the magnetic and electric helicities, as we have seen in the previous section. These results have been obtained in the framework of standard classical electromagnetism in a vacuum, but they are also compatible with the suggestion made by Bliokh of a dual theory of electromagnetism [9].

6. The Angular Momentum Decomposition for Non-Null Toroidal Electromagnetic Fields

In this section, we calculate explicitly and analytically the spin-angular decomposition of a whole class of electromagnetic fields in a vacuum without using any paraxial approximation.

We will use the knotted non-null torus class [5,25]. These fields are exact solutions of Maxwell equations in a vacuum with the property that, at a given time $t = 0$, all pairs of lines of the field $\mathbf{B}(\mathbf{r}, 0)$ are linked torus knots and that the linking number is the same for all the pairs. Similarly, for the electric field at the initial time $\mathbf{E}(\mathbf{r}, 0)$, all pairs of lines are linked torus knots, and the linking number is the same for all the pairs.

We take a four positive integers tuplet (n, m, l, s). It is possible to find an initial magnetic field such that all its magnetic lines are (n, m) torus knots. The linking number of every two magnetic lines at $t = 0$ is equal to nm. Furthermore, we can find an initial electric field such that all the electric lines are (l, s) torus knots and at $t = 0$. At that time, the linking number of the electric field lines is equal to ls. We can assure that property at $t = 0$, due to the fact that the topology may change during time evolution if one of the integers (n, m, l, s) is different from any of the others (for details, we refer the interested reader to [5]). The magnetic and electric helicities also may change if the integer tuplet is not proportional to (n, n, l, l). In these cases, the electromagnetic fields interchange the magnetic and electric helicities during their time evolution.

We define the dimensionless coordinates (X, Y, Z, T), which are related to the physical ones (x, y, z, t) by $(X, Y, Z, T) = (x, y, z, ct)/L_0$, and $r^2/L_0^2 = (x^2 + y^2 + z^2)/L_0^2 = X^2 + Y^2 + Z^2 = R^2$. The length scale L_0 can be chosen to be the mean quadratic radius of the energy distribution of the electromagnetic field. The set of non-null torus electromagnetic knots can be written as:

$$\mathbf{B}(\mathbf{r}, t) = \frac{\sqrt{a}}{\pi L_0^2} \frac{Q\mathbf{H}_1 + P\mathbf{H}_2}{(A^2 + T^2)^3} \tag{53}$$

$$\mathbf{E}(\mathbf{r}, t) = \frac{\sqrt{a}c}{\pi L_0^2} \frac{Q\mathbf{H}_4 - P\mathbf{H}_3}{(A^2 + T^2)^3} \tag{54}$$

where a is a constant related to the energy of the electromagnetic field,

$$A = \frac{1 + R^2 - T^2}{2}, \quad P = T(T^2 - 3A^2), \quad Q = A(A^2 - 3T^2), \tag{55}$$

and:

$$\mathbf{H}_1 = (-nXZ + mY + sT)\mathbf{u}_x + (-nYZ - mX - lTZ)\mathbf{u}_y \\ + \left(n\frac{-1 - Z^2 + X^2 + Y^2 + T^2}{2} + lTY\right)\mathbf{u}_z. \tag{56}$$

$$\mathbf{H}_2 = \left(s\frac{1 + X^2 - Y^2 - Z^2 - T^2}{2} - mTY\right)\mathbf{u}_x + (sXY - lZ + mTX)\mathbf{u}_y + (sXZ + lY + nT)\mathbf{u}_z. \tag{57}$$

$$\mathbf{H}_3 = (-mXZ + nY + lT)\mathbf{u}_x + (-mYZ - nX - sTZ)\mathbf{u}_y \\ + \left(m\frac{-1 - Z^2 + X^2 + Y^2 + T^2}{2} + sTY\right)\mathbf{u}_z. \tag{58}$$

$$\mathbf{H}_4 = \left(l\frac{1 + X^2 - Y^2 - Z^2 - T^2}{2} - nTY\right)\mathbf{u}_x + (lXY - sZ + nTX)\mathbf{u}_y + (lXZ + sY + mT)\mathbf{u}_z. \tag{59}$$

The energy $\mathcal{E}$, linear momentum $\mathbf{p}$ and total angular momentum $\mathbf{J}$ of these fields are:

$$\mathcal{E} = \int \left(\frac{\epsilon_0 E^2}{2} + \frac{B^2}{2\mu_0}\right) d^3r = \frac{a}{2\mu_0 L_0}(n^2 + m^2 + l^2 + s^2) \tag{60}$$

$$\mathbf{p} = \int \epsilon_0 \mathbf{E} \times \mathbf{B}\, d^3r = \frac{a}{2c\mu_0 L_0}(ln + ms)\mathbf{u}_y \tag{61}$$

$$\mathbf{J} = \int \epsilon_0 \mathbf{r} \times (\mathbf{E} \times \mathbf{B})\, d^3r = \frac{a}{2c\mu_0}(lm + ns)\mathbf{u}_y \tag{62}$$

To study the interchange between the magnetic and electric helicities and the spins, we first need the Fourier transforms of the fields in the helicity basis. Following the prescription given in Section 3, we get:

$$
\begin{aligned}
a_R \mathbf{e}_R + a_L \mathbf{e}_L &= \sqrt{\frac{a}{\hbar c \mu_0}} \frac{L_0^{3/2}}{2\sqrt{\pi}} \frac{e^{-K}}{\sqrt{K}} \times \left[\frac{m}{K} \left(K_x K_z, K_y K_z, -K_x^2 - K_y^2 \right) + s \left(0, K_z, -K_y \right) \right] \\
&+ i \left[\frac{l}{K} \left(-K_y^2 - K_z^2, K_x K_y, K_x K_z \right) + n \left(-K_y, K_x, 0 \right) \right]
\end{aligned}
\tag{63}
$$

$$
\begin{aligned}
a_R \mathbf{e}_R - a_L \mathbf{e}_L &= \sqrt{\frac{a}{\hbar c \mu_0}} \frac{L_0^{3/2}}{2\sqrt{\pi}} \frac{e^{-K}}{\sqrt{K}} \times \left[\frac{n}{K} \left(K_x K_z, K_y K_z, -K_x^2 - K_y^2 \right) + l \left(0, K_z, -K_y \right) \right] \\
&+ i \left[\frac{s}{K} \left(-K_y^2 - K_z^2, K_x K_y, K_x K_z \right) + m \left(-K_y, K_x, 0 \right) \right]
\end{aligned}
\tag{64}
$$

In these expressions, we have introduced the dimensionless Fourier space coordinates (K_x, K_y, K_z), related to the dimensional Fourier space coordinates (k_x, k_y, k_z) according to:

$$
(K_x, K_y, K_z) = L_0(k_x, k_y, k_z), \quad K = L_0 k = \frac{L_0 \omega}{c}.
\tag{65}
$$

The electromagnetic helicity Equation (37) of the set of non-null torus electromagnetic knots results:

$$
h = \hbar \int d^3k \, (\bar{a}_R(\mathbf{k}) a_R(\mathbf{k}) - \bar{a}_L(\mathbf{k}) a_L(\mathbf{k})) = \frac{a}{2c\mu_0}(nm + ls),
\tag{66}
$$

and the difference between the magnetic and electric helicities is:

$$
\begin{aligned}
\tilde{h}(t) &= h_m - h_e = \frac{\hbar}{2} \int d^3k \left[e^{-2i\omega t} \left(-a_R(\mathbf{k}) a_R(-\mathbf{k}) + a_L(\mathbf{k}) a_L(-\mathbf{k}) \right) \right. \\
&\left. + e^{2i\omega t} \left(-\bar{a}_R(\mathbf{k}) \bar{a}_R(-\mathbf{k}) + \bar{a}_L(\mathbf{k}) \bar{a}_L(-\mathbf{k}) \right) \right] \\
&= \frac{a}{2c\mu_0}(nm - ls) \frac{1 - 6T^2 + T^4}{(1 + T^2)^4},
\end{aligned}
\tag{67}
$$

where we recall that $T = ct/L_0$. Results Equations (66) and (67) coincide with the computations done in [5] using different procedures.

Now, consider the spin in Equation (48). For the set of non-null torus electromagnetic knots, we get:

$$
\mathbf{s} = \hbar \int d^3k \, (\bar{a}_R(\mathbf{k}) a_R(\mathbf{k}) - \bar{a}_L(\mathbf{k}) a_L(\mathbf{k})) \, \mathbf{e}_k = \frac{a}{4c\mu_0}(ml + ns) \, \mathbf{u}_y.
\tag{68}
$$

Notice that this value of spin is equal to one half of the value of the total angular momentum obtained in Equation (62). Thus, the orbital angular momentum of this set of electromagnetic fields has the same value as the spin angular momentum,

$$
\mathbf{L} = \mathbf{s} = \frac{1}{2} \mathbf{J}.
\tag{69}
$$

The difference between the magnetic and the electric spin can also be computed through Equation (49). The result is:

$$
\begin{aligned}
\tilde{\mathbf{s}}(t) &= \mathbf{s}_m - \mathbf{s}_e = \frac{\hbar}{2} \int d^3k \left[e^{-2i\omega t} \left(a_R(\mathbf{k}) a_R(-\mathbf{k}) - a_L(\mathbf{k}) a_L(-\mathbf{k}) \right) \right. \\
&\left. + e^{2i\omega t} \left(\bar{a}_R(\mathbf{k}) \bar{a}_R(-\mathbf{k}) - \bar{a}_L(\mathbf{k}) \bar{a}_L(-\mathbf{k}) \right) \right] \mathbf{e}_k = 0.
\end{aligned}
\tag{70}
$$

As a consequence, even if the magnetic and electric helicities depend on time for this set of electromagnetic fields, the magnetic and electric parts of the spin are time independent, satisfying the results found in Equation (52) for general electromagnetic fields in a vacuum. Both are equal, and satisfy:

$$\mathbf{s}_m = \mathbf{s}_e = \frac{1}{2}\mathbf{s} = \frac{a}{8c\mu_0}(ml + ns)\,\mathbf{u}_y. \tag{71}$$

7. Same Spin-Orbital Decomposition with Different Behavior in the Helicities

In this section, we consider two knotted electromagnetic fields in which the spin and orbital decomposition of the angular momentum are equal in both cases, while the helicities are constant and non-constant, respectively. We will see that the angular momentum density evolves differently in each case.

In the first case, we take the set $(n, m, l, s) = (5, 3, 5, 3)$ in Equations (53) and (54). Thus, using Equation (62) the total angular momentum is:

$$\mathbf{J} = \frac{15a}{\mu_0}\,\mathbf{u}_y,$$

while the angular density changes in time. In order to visualize the evolution of the angular momentum density, which is given by $\mathbf{j} = \mathbf{r} \times (\mathbf{E} \times \mathbf{B})$, we plot at different times the vector field sample at the plane XZ, as is depicted in Figure 1.

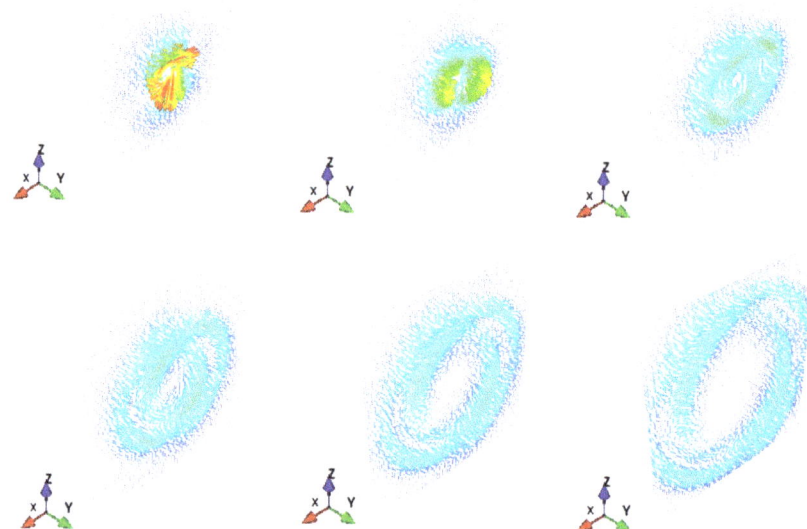

Figure 1. The angular momentum density $\mathbf{j}$ at times $T = 0, 0.5, 1, 1.5, 2, 2.5$, for the electromagnetic field given by the set $(n, m, l, s) = (5, 3, 5, 3)$. The vector field is sampled at the plane XZ. In the case depicted in the figure, the magnetic helicity is equal to the electric helicity and constant in time.

For this case, the spin-orbital split, as shown in the previous section, using Equation (69), turns out to be:

$$\mathbf{L} = \mathbf{s} = \frac{15a}{2\mu_0}\,\mathbf{u}_y. \tag{72}$$

which remains constant during the time evolution of the field. The magnetic and electric helicities remain also constant, and there is no exchange between them.

Now, let us take the set $(n, m, l, s) = (15, 5, 0, 2)$ in Equations (53) and (54). The electromagnetic field obtained with this set of integers has the same value of the total angular momentum as the previous case and the same spin-orbital split. However, in this case, the magnetic and electric helicities are time-dependent, satisfying Equation (67). The time evolution of the angular momentum density is different from the case of constant helicities, as we can see in Figure 2. As we did before, we have plotted the field at the plane XZ at the same time steps as in the first example.

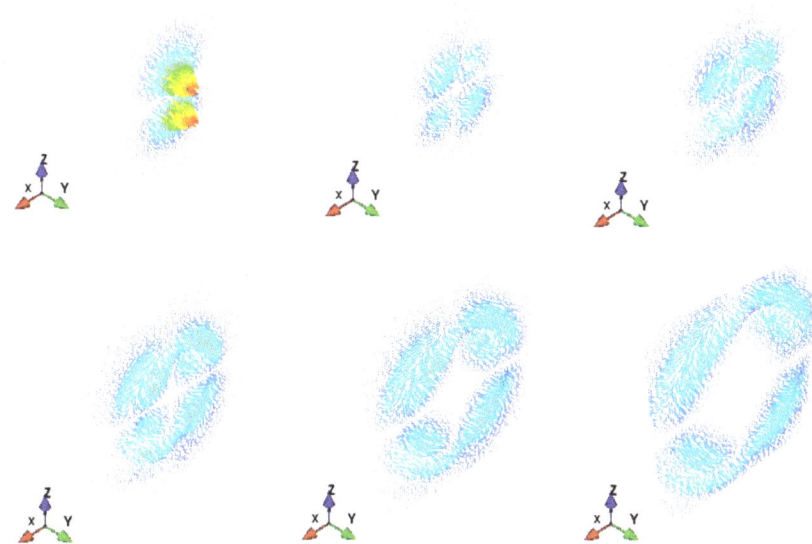

Figure 2. The angular momentum density **j** at times $T = 0, 0.5, 1, 1.5, 2, 2.5$, for the electromagnetic field given by the set $(n, m, l, s) = (15, 5, 0, 2)$. The vector field is sampled at the plane XZ. In this example, the magnetic and electric helicities are initially different, and their values change with time.

In the first example of a non-null torus electromagnetic field, the helicities remain constant in time. In the second example, the magnetic helicity is initially different from the electric helicity, and both change with time. Even if the spin, orbital and total angular momenta are equal in both examples, we can see in Figures 1 and 2 that the structure of the total angular momentum density is different. We can speculate that a macroscopic particle, which can interact with the angular momentum of the field, would behave in the same way in both cases, but a microscopic test particle able to interact with the local density of the angular momentum would behave differently.

8. Conclusions

We have calculated analytically and exactly the spin-orbital decomposition of the angular momentum of a class of electromagnetic fields beyond the paraxial approximation. A spin density that is dual in its magnetic and electric contributions has been considered. This spin density has the meaning of flux of electromagnetic helicity. By using a Fourier decomposition of the electromagnetic field in a vacuum in terms of circularly-polarized waves, called the helicity basis, we have given explicit expressions for the magnetic and electric contributions to the spin angular momentum. We have obtained the results that the magnetic and electrical components of spin remain constant during the time evolution of the fields. We also have made use of the helicity basis to calculate the magnetic and electric helicities.

We have obtained the exact split of the angular momentum into spin and orbital components for electromagnetic fields, which belong to the non-null toroidal knotted class [5]. One of main characteristics of that class is that it contains a certain degree of linkage of electric and magnetic lines and can have exchange between the magnetic and electrical components of the helicity [8].

We have considered two examples of these non-null knotted electromagnetic fields having the properties that they have the same angular momentum and the same split. They have the same constant values for the orbital and spin components of the angular momentum, the first with constant and equal helicities and the second with time-evolving helicities. The behavior of the total angular momentum density seems to be different in these two cases.

In our opinion, the study of this kind of example with nontrivial helicities may provide a clarification of the role of helicities in the behavior of angular momentum densities of electromagnetic fields in a vacuum.

Acknowledgments: We acknowledge Wolfgang Löffler for valuable discussions. This work was supported by research grants from the Spanish Ministry of Economy and Competitiveness (MINECO/FEDER) ESP2015-69909-C5-4-R.

Author Contributions: Manuel Arrayás and José L. Trueba conceived of all the results of this work, made the computations and wrote the paper.

Conflicts of Interest: The authors declare no conflict of interest. The founding sponsors had no role in the design of the study; in the collection, analyses or interpretation of data; in the writing of the manuscript; nor in the decision to publish the results.

References

1. Van Enk, S.J.; Nienhuis, G. Spin and orbital angular momentum of photons. *Europhys. Lett.* **1994**, *25*, 497–501.
2. Van Enk, S.J.; Nienhuis, G. Commutation rules and eigenvalues of spin and orbital angular momentum of radiation fields. *J. Mod. Opt.* **1994**, *41*, 963–977.
3. Allen, L.; Barnett, S.M.; Padgett, M.J. (Eds.) *Optical Angular Momentum*; Institute of Physics: Bristol, UK, 2003.
4. Bliokh, K.Y.; Aiello, A.; Alonso, M. *The Angular Momentum of Light*; Andrews, D.L., Babiker, M., Eds; Cambridge University Press: Hong Kong, China, 2012.
5. Arrayás, M.; Trueba, J.L. A class of non-null toroidal electromagnetic fields and its relation to the model of electromagnetic knots. *J. Phys. A Math. Theor.* **2015**, *48*, 025203.
6. Afanasiev, G.N.; Stepanovsky, Y.P. The helicity of the free electromagnetic field and its physical meaning. *Nuovo Cim. A* **1996**, *109*, 271–279.
7. Trueba, J.L.; Rañada, A.F. The electromagnetic helicity. *Eur. J. Phys.* **1996**, *17*, 141–144.
8. Arrayás, M.; Trueba, J.L. Exchange of helicity in a knotted electromagnetic field. *Ann. Phys. (Berl.)* **2012**, *524*, 71–75.
9. Bliokh, K.Y.; Bekshaev, A.Y.; Nori, F. Dual electromagnetism: Helicity, spin, momentum and angular momentum. *New J. Phys.* **2013**, *15*, 033026.
10. Berry, M.V. Optical currents. *J. Opt. A Pure Appl. Opt.* **2009**, *11*, 094001.
11. Ra nada, A.F. A topological theory of the electromagnetic field. *Lett. Math. Phys.* **1989**, *18*, 97–106.
12. Ra nada, A.F. Knotted solutions of the Maxwell equations in a vacuum. *J. Phys. A Math. Gen.* **1990**, *23*, L815–L820.
13. Ra nada, A.F. Topological electromagnetism. *J. Phys. A Math. Gen.* **1992**, *25*, 1621–1641.
14. Ra nada, A.F.; Trueba, J.L. Electromagnetic knots. *Phys. Lett. A* **1995**, *202*, 337–342.
15. Ra nada, A.F.; Trueba, J.L. Two properties of electromagnetic knots. *Phys. Lett. A* **1997**, *232*, 25–33.
16. Ra nada, A.F.; Trueba, J.L. A topological mechanism of discretization for the electric charge. *Phys. Lett. B* **1998**, *422*, 196–200.
17. Ra nada, A.F.; Trueba, J.L. Topological Electromagnetism with Hidden Nonlinearity. In *Modern Nonlinear Optics III*; Evans, M.W., Ed.; John Wiley & Sons: New York, NY, USA, 2001; pp 197–253.
18. Irvine, W.T.M.; Bouwmeester, D. Linked and knotted beams of light. *Nat. Phys.* **2008**, *4*, 716–720.

19. Besieris, I.M.; Shaarawi, A.M. Hopf-Rañada linked and knotted light beam solution viewed as a null electromagnetic field. *Opt. Lett.* **2009**, *34*, 3887–3889.
20. Arrayás, M.; Trueba, J.L. Motion of charged particles in a knotted electromagnetic field. *J. Phys. A Math. Theor.* **2010**, *43*, 235401.
21. Van Enk, S.J. The covariant description of electric and magnetic field lines of null fields: Application to Hopf-Rañada solutions. *J. Phys. A Math. Theor.* **2013**, *46*, 175204.
22. Kedia, H.; Bialynicki-Birula, I.; Peralta-Salas, D.; Irvine, W.T.M. Tying knots in light fields. *Phys. Rev. Lett.* **2013**, *111*, 150404.
23. Hoyos, C.; Sircar, N.; Sonnenschein, J. New knotted solutions of Maxwell's equations. *J. Phys. A Math. Theor.* **2015**, *48*, 255204.
24. Kedia, H.; Foster, D.; Dennis, M.R.; Irvine, W.T.M. Weaving knotted vector fields with tunable helicity. *Phys. Rev. Lett.* **2016**, *117*, 274501.
25. Arrayás, M.; Bouwmeester, D.; Trueba, J.L. Knots in electromagnetism. *Phys. Rep.* **2017**, *667*, 1–61.
26. Arrayás, M.; Trueba, J.L. Collision of two hopfions. *J. Phys. A Math. Theor.* **2017**, *50*, 085203.
27. Kamchatnov, A.M. Topological solitons in magnetohydrodynamics. *Zh. Eksp. Teor. Fiz.* **1982**, *82*, 117–124.
28. Semenov, V.S.; Korovinski, D.B.; Biernat, H.K. Euler potentials for the MHD Kamchatnov-Hopf soliton solution. *Nonlinear Process. Geophys.* **2002**, *9*, 347–354.
29. Thompson, A.; Swearngin, J.; Wickes, A.; Bouwmeester, D. Constructing a class of topological solitons in magnetohydrodynamics. *Phys. Rev. E* **2014**, *89*, 043104.
30. Smiet, C.B.; Candelaresi, S.; Thompson, A.; Swearngin, J.; Dalhuisen, J.W.; Bouwmeester, D. Self-organizing knotted magnetic structures in plasma. *Phys. Rev. Lett.* **2015**, *115*, 095001.
31. O'Holleran, K.; Dennis, M.R.; Padgett, M.J. Topology of light's darkness. *Phys. Rev. Lett.* **2009**, *102*, 143902.
32. Dennis, M.R.; King, R.P.; Jack, B.; O'Holleran, K.; Padgett, M.J. Isolated optical vortex knots. *Nat. Phys.* **2010**, *6*, 118–121.
33. Romero, J.; Leach, J.; Jack, B.; Dennis, M.R.; Franke-Arnold, S.; Barnett, S.M.; Padgett, M.J. Entangled optical vortex links. *Phys. Rev. Lett.* **2011**, *106*, 100407.
34. Desyatnikov, A.S.; Buccoliero, D.; Dennis, M.R.; Kivshar, Y.S. Spontaneous knotting of self-trapped waves. *Sci. Rep.* **2012**, *2*, 771.
35. Rubinsztein-Dunlop, H.; Forbes, A.; Berry, M.V.; Dennis, M.R.; Andrews, D.L.; Mansuripur, M.; Denz, C.; Alpmann, C.; Banzer, P.; Bauer, T.; et al. Roadmap on Structured Light. *J. Opt.* **2017**, *19*, 013001.
36. Faddeev, L.; Niemi, A.J. Stable knot-like structures in classical field theory. *Nature* **1997**, *387*, 58–61.
37. Hall, D.S.; Ray, M.W.; Tiurev, K.; Ruokokoski, E.; Gheorge, A.H.; Möttönen, M. Tying quantum knots. *Nat. Phys.* **2016**, *12*, 478–483.
38. Taylor, A.J.; Dennis, M.R. Vortex knots in tangled quantum eigenfunctions. *Nat. Commun.* **2016**, *7*, 12346.
39. Volovik, G.E.; Mineev, V.O. Particle-like solitons in superfluid He phases. *Zh. Eksp. Teor. Fiz.* **1977**, *73*, 767–773.
40. Dzyloshinskii, I.; Ivanov, B. Localized topological solitons in a ferromagnet. *Pis'ma Zh. Eksp. Teor. Fiz.* **1979**, *29*, 592–595.
41. Kawaguchi, Y.; Nitta, M.; Ueda, M. Knots in a spinor Bose-Einstein condensate. *Phys. Rev. Lett.* **2008**, *100*, 180403.
42. Kleckner, A.; Irvine, W.T.M. Creation and dynamics of knotted vortices. *Nat. Phys.* **2013**, *9*, 253–258.
43. Kleckner, A.; Irvine, W.T.M. Liquid crystals: Tangled loops and knots. *Nat. Mat.* **2014**, *13*, 229–231.
44. Dalhuisen, J.W.; Bouwmeester, D. Twistors and electromagnetic knots. *J. Phys. A Math. Theor.* **2012**, *45*, 135201.
45. Thompson, A.; Swearngin, J.; Wickes, A.; Bouwmeester, D. Classification of electromagnetic and gravitational hopfions by algebraic type. *J. Phys. A Math. Theor.* **2015**, *48*, 205202.
46. Barnett, S.M. On the six components of optical angular momentum. *J. Opt.* **2011**, *13*, 064010.
47. Stratton, J.A. *Electromagnetic Theory*; McGraw-Hill: New York, NY, USA, 1941.
48. Moffatt, H.K. The degree of knottedness of tangled vortex lines. *J. Fluid Mech.* **1969**, *35*, 117–129.
49. Berger, M.A.; Field, G.B. The topological properties of magnetic helicity. *J. Fluid Mech.* **1984**, *147*, 133–148.
50. Moffatt, H.K.; Ricca, R.L. Helicity and the Calugareanu Invariant. *Proc. R. Soc. A* **1992**, *439*, 411–429.
51. Berger, M.A. Introduction to magnetic helicity. *Plasma Phys. Control. Fusion* **1999**, *41*, B167–B175.
52. Dennis, M.R.; Hannay, J.H. Geometry of Calugareanu's theorem. *Proc. R. Soc. A* **2005**, *461*, 3245–3254.
53. Ricca, R.L.; Nipoti, B. Gauss' linking number revisited. *J. Knot Theor. Ramif.* **2011**, *20*, 1325–1343.
54. Bliokh, K.Y.; Alonso, M.A.; Ostrovskaya, E.A.; Aiello, A. Angular momenta and spin-orbit interaction of nonparaxial light in free space. *Phys. Rev. A* **2010**, *82*, 063825.

55. Barnett, S.M. Rotation of electromagnetic fields and the nature of optical angular momentum. *J. Mod. Opt.* **2010**, *57*, 1339–1343.
56. Bialynicki-Birula, I.; Bialynicki-Birula, Z. Canonical separation of angular momentum of light into its orbital and spin parts. *J. Opt.* **2011**, *13*, 064014.
57. Ynduráin, F.J. *Mecánica Cuántica*; Alianza Editorial: Madrid, Spain, 1988.
58. Bialynicki-Birula, I. Local and nonlocal observables in quantum optics. *New J. Phys.* **2014**, *16*, 113056.

 © 2018 by the authors. Licensee MDPI, Basel, Switzerland. This article is an open access article distributed under the terms and conditions of the Creative Commons Attribution (CC BY) license (http://creativecommons.org/licenses/by/4.0/).

Article

Null Electromagnetic Fields from Dilatation and Rotation Transformations of the Hopfion

Manuel Arrayás [1,†], Antonio F. Rañada [2,†], Alfredo Tiemblo [3,†] and José L. Trueba [1,*,†]

1. Área de Electromagnetismo, Universidad Rey Juan Carlos, Calle Tulipán s/n, 28933 Móstoles (Madrid), Spain; manuel.arrayas@urjc.es
2. Departamento de Física Aplicada III, Universidad Complutense, Plaza de las Ciencias s/n, 28040 Madrid, Spain; antonio.f.ranada@gmail.com
3. Instituto de Física Fundamental, Consejo Superior de Investigaciones Científicas, Calle Serrano 113, 28006 Madrid, Spain; alfredotiemblo@gmail.com
* Correspondence: joseluis.trueba@urjc.es
† These authors contributed equally to this work.

Received: 22 July 2019; Accepted: 28 August 2019; Published: 2 September 2019

Abstract: The application of topology concepts to Maxwell equations has led to the developing of the whole area of electromagnetic knots. In this paper, we apply some symmetry transformations to a particular electromagnetic knot, the hopfion field, to get a new set of knotted solutions with the properties of being null. The new fields are obtained by a homothetic transformation (dilatation) and a rotation of the hopfion, and we study the constraints that the transformations must fulfill in order to generate valid electromagnetic fields propagating in a vacuum. We make use of the Bateman construction and calculate the four-potentials and the electromagnetic helicities. It is observed that the topology of the field lines does not seem to be conserved as it is for the hopfion.

Keywords: hopfion; Bateman construction; null fields

---

1. Introduction

In recent years, topology ideas applied to physics have provided useful insights into many different phenomena, ranging from phase transitions to solid state physics, superfluids, and magnetism. The topology applied to electromagnetism has opened the field of electromagnetic knots [1], where light gets nontrivial properties. One example of a electromagnetic knot is the hopfion.

The hopfion is an exact null solution of the Maxwell equations in a vacuum [1–4]. The null property means that the Lorentz invariants of the field are zero, i.e., $\mathbf{E} \cdot \mathbf{B} = 0$ and $E^2 - c^2 B^2 = 0$ [5]. The hopfion is characterized by further special properties such as the field lines being closed and linked for any instant of time. The topology of the field lines is described for any time in terms of two complex scalar fields $\phi(\mathbf{r}, t)$ and $\theta(\mathbf{r}, t)$. The solution of $\phi(\mathbf{r}, t) = c_1$ and $\theta(\mathbf{r}, t) = c_2$, with $c_1, c_2 \in \mathcal{C}$ complex constants, gives all the magnetic and electric lines (by changing the value of the constant at the right-hand side), which are linked closed lines topologically equivalent to circles. In particular, for the hopfion, those complex fields can be written in terms of four real scalar fields u_1, u_2, u_3, u_4, as:

$$\phi = \frac{u_1 + i\, u_2}{u_3 + i\, u_4}, \tag{1}$$

$$\theta = \frac{u_2 + i\, u_3}{u_1 + i\, u_4}, \tag{2}$$

The u_i's satisfy the conditions $-1 \leq u_i \leq 1$ and $u_1^2 + u_2^2 + u_3^2 + u_4^2 = 1$, so they can be considered as time-dependent coordinates on the sphere S^3. In this case, the ϕ and θ are then applications from $S^3 \to S^2$, and the linking properties of the field lines follow from this fact [1].

In this paper, we apply some symmetry transformations to the hopfion. In particular, we make a homothetic transformation (dilatation) and a rotation of the hopfion at a particular time. We find that those transformations cannot be arbitrary in order for the transformed fields to be still electromagnetic solutions. We provide the conditions required for the transformations. We give explicit expressions for the new null fields for any time using the Bateman construction. Furthermore, the four-potentials of the new fields and the electromagnetic helicities are calculated. The non-null helicities point to the fact that the topology of the field lines is not trivial. However, the new solutions do not seem to preserve the closedness property of the hopfion field lines. This fact deserves future investigations.

2. Topological Construction of Vacuum Solutions and the Hopfion Field

In this section, we will briefly revise a topological formulation of electromagnetism in a vacuum built in [3,6–8] and give the explicit expression for the hopfion field. We will make use of certain properties of this construction in the next section when we apply the transformations to the hopfion.

Solutions of Maxwell equations in a vacuum (we will use MKSunits),

$$\nabla \cdot \mathbf{B} = 0, \tag{3}$$

$$\nabla \times \mathbf{E} = -\frac{\partial \mathbf{B}}{\partial t}, \tag{4}$$

$$\nabla \cdot \mathbf{E} = 0, \tag{5}$$

$$\nabla \times \mathbf{B} = \frac{1}{c^2}\frac{\partial \mathbf{E}}{\partial t}, \tag{6}$$

can be found from a pair of complex scalar fields $\phi(\mathbf{r},t)$ and $\theta(\mathbf{r},t)$, so the magnetic end electric fields are given by:

$$\mathbf{B} = \frac{\sqrt{a}}{2\pi i}\frac{\nabla\phi \times \nabla\bar\phi}{(1+\phi\bar\phi)^2} = \frac{\sqrt{a}}{2\pi i c}\frac{\partial_t\bar\theta\nabla\theta - \partial_t\theta\nabla\bar\theta}{(1+\theta\bar\theta)^2}, \tag{7}$$

$$\mathbf{E} = \frac{\sqrt{ac}}{2\pi i}\frac{\nabla\bar\theta \times \nabla\theta}{(1+\theta\bar\theta)^2} = \frac{\sqrt{a}}{2\pi i}\frac{\partial_t\bar\phi\nabla\phi - \partial_t\phi\nabla\bar\phi}{(1+\phi\bar\phi)^2}, \tag{8}$$

As usual, c denotes the speed of light, and a is a constant so that the magnetic and electric fields have correct dimensions in MKS units since ϕ and θ are dimensionless ($\bar\phi$ is the complex conjugate of ϕ). Equation (3) follows from the first equality of Equation (7). Equation (4) is found using the first equality of Equation (7) and the second equality of Equation (8). Equation (5) comes from the first equality of Equation (8). Equation (6) is fulfilled considering the second equality of Equation (7) and the first equality of Equation (8).

To get a solution of Maxwell equations in a vacuum, the complex scalar fields $\phi(\mathbf{r},t)$ and $\theta(\mathbf{r},t)$ have to be found to satisfy Equations (7) and (8), so:

$$\frac{\nabla\phi \times \nabla\bar\phi}{(1+\phi\bar\phi)^2} = \frac{1}{c}\frac{\partial_t\bar\theta\nabla\theta - \partial_t\theta\nabla\bar\theta}{(1+\theta\bar\theta)^2}, \tag{9}$$

$$\frac{\nabla\bar\theta \times \nabla\theta}{(1+\theta\bar\theta)^2} = \frac{1}{c}\frac{\partial_t\bar\phi\nabla\phi - \partial_t\phi\nabla\bar\phi}{(1+\phi\bar\phi)^2}. \tag{10}$$

These equations are a bit cumbersome, although some solutions have been found in the literature [2,3,9]. On the other side, the advantage of this formulation is that the magnetic and the electric lines are very easily obtained. The field lines at a given time t correspond to the level curves of the scalar field $\phi(\mathbf{r},t)$ and $\theta(\mathbf{r},t)$. This observation can be particularly useful to find solutions of Maxwell equations in a vacuum in which the magnetic and electric lines form knotted curves [1,3]. In this case, the degree of knottedness has interesting physical consequences [4,10–16].

All the solutions of Maxwell equations in this particular formulation satisfy the Lorentz-invariant equation $\mathbf{E} \cdot \mathbf{B} = 0$. This can be immediately seen by using the first equality of Equation (7) and the second equality of Equation (8) or, correspondingly, the second equality of Equation (7) and the first equality of Equation (8). However, it is not true that all the solutions in this formulation satisfy the other null condition, $E^2 - c^2 B^2 = 0$.

The hopfion was found choosing the particular form Equation (1) for ϕ and Equation (2) for θ. In terms of the four real scalar fields u_1, u_2, u_3, u_4, using Equations (7) and (8) turns out to be:

$$\mathbf{B}_H(\mathbf{r},t) = -\frac{\sqrt{a}}{\pi} (\nabla u_1 \times \nabla u_2 + \nabla u_3 \times \nabla u_4)$$
$$= \frac{\sqrt{a}}{\pi c} \left(\frac{\partial u_2}{\partial t} \nabla u_3 - \frac{\partial u_3}{\partial t} \nabla u_2 + \frac{\partial u_1}{\partial t} \nabla u_4 - \frac{\partial u_4}{\partial t} \nabla u_1 \right), \quad (11)$$

$$\mathbf{E}_H(\mathbf{r},t) = \frac{c\sqrt{a}}{\pi} (\nabla u_2 \times \nabla u_3 + \nabla u_1 \times \nabla u_4)$$
$$= \frac{\sqrt{a}}{\pi} \left(\frac{\partial u_1}{\partial t} \nabla u_2 - \frac{\partial u_2}{\partial t} \nabla u_1 + \frac{\partial u_3}{\partial t} \nabla u_4 - \frac{\partial u_4}{\partial t} \nabla u_3 \right). \quad (12)$$

The explicit expressions for the u_i's are:

$$u_1 = \frac{AX - TZ}{A^2 + T^2}, \quad (13)$$

$$u_2 = \frac{AY + T(A-1)}{A^2 + T^2}, \quad (14)$$

$$u_3 = \frac{AZ + TX}{A^2 + T^2}, \quad (15)$$

$$u_4 = \frac{A(A-1) - TY}{A^2 + T^2}. \quad (16)$$

where:

$$A = \frac{R^2 - T^2 + 1}{2}, \quad R^2 = X^2 + Y^2 + Z^2, \quad (17)$$

and (X, Y, Z, T) are dimensionless coordinates. Spacetime coordinates (x, y, z, t) are related to them as:

$$(x, y, z, t) = (L_0 X, L_0 Y, L_0 Z, L_0 T/c), \quad (18)$$

L_0 being a constant with length dimensions, which characterizes the mean quadratic radius of the electromagnetic energy distribution [17].

It is easy to see, given Expressions (13)–(16), that the u_i's satisfy the conditions $-1 \leq u_i \leq 1$ and $u_1^2 + u_2^2 + u_3^2 + u_4^2 = 1$ for any time. For the hopfion, both null conditions $\mathbf{E} \cdot \mathbf{B} = 0$ and $E^2 - c^2 B^2 = 0$ are satisfied.

3. Dilatation and Rotation of the Hopfion

In this section, we explore the possibility of obtaining new solutions by symmetry transformations. We will apply a family of transformations to the hopfion: a dilatation and a rotation at a particular time. We then check the conditions imposed by the initial conditions of Maxwell solutions to find that the transformations must fulfill some constraints expressed as differential equations. The equations are then solved in order to determine a particular set of allowed transformations. In the next section, we will extend the results to every time and generate a more general transformation of the hopfion field that satisfies Maxwell equations in a vacuum.

A warning: along this section, the notation is simplified by taking $a = 1$, $L_0 = 1$, $c = 1$, and we will write coordinates X, Y, Z, T, R as x, y, z, t, r, respectively, in all the computations. However, the final results will be written back with all the constants, so that they can be used in different contexts.

The particular time $t = 0$ is chosen to apply the transformations as the expressions are simpler. The hopfion at this particular time can be written using the new notation as:

$$\mathbf{B}_{H,0}(\mathbf{r}) = \frac{8}{\pi(r^2+1)^3} \mathbf{e}_1,$$
$$\mathbf{E}_{H,0}(\mathbf{r}) = \frac{8}{\pi(r^2+1)^3} \mathbf{e}_2, \quad (19)$$

and the Poynting vector $\mathbf{P} = \mathbf{E} \times \mathbf{B}$ reads:

$$\mathbf{P}_{H,0}(\mathbf{r}) = -\frac{64}{\pi^2(r^2+1)^5} \mathbf{e}_3, \quad (20)$$

where the vector fields:

$$\mathbf{e}_1 = \left(y - xz, -x - yz, \frac{x^2+y^2-z^2-1}{2}\right),$$
$$\mathbf{e}_2 = \left(\frac{x^2-y^2-z^2+1}{2}, -z+xy, y+xz\right),$$
$$\mathbf{e}_3 = \left(-z - xy, \frac{x^2-y^2+z^2-1}{2}, x - yz\right). \quad (21)$$

have been defined. They constitute a basis in the three-dimensional Euclidean space, satisfying:

$$\mathbf{e}_1 \cdot \mathbf{e}_1 = \mathbf{e}_2 \cdot \mathbf{e}_2 = \mathbf{e}_3 \cdot \mathbf{e}_3 = \left(\frac{r^2+1}{2}\right)^2,$$
$$\mathbf{e}_1 \cdot \mathbf{e}_2 = \mathbf{e}_2 \cdot \mathbf{e}_3 = \mathbf{e}_3 \cdot \mathbf{e}_1 = 0,$$
$$\mathbf{e}_1 \times \mathbf{e}_2 = \left(\frac{r^2+1}{2}\right) \mathbf{e}_3,$$
$$\mathbf{e}_2 \times \mathbf{e}_3 = \left(\frac{r^2+1}{2}\right) \mathbf{e}_1,$$
$$\mathbf{e}_3 \times \mathbf{e}_1 = \left(\frac{r^2+1}{2}\right) \mathbf{e}_2. \quad (22)$$

We will make, as stated above, a dilatation and rotation of the fields and write the transformed fields as:

$$\mathbf{B}_0(\mathbf{r}) = f(r^2)\left(\cos\eta\, \mathbf{B}_{H,0} + \sin\eta\, \mathbf{E}_{H,0}\right),$$
$$\mathbf{E}_0(\mathbf{r}) = f(r^2)\left(-\sin\eta\, \mathbf{B}_{H,0} + \cos\eta\, \mathbf{E}_{H,0}\right), \quad (23)$$

where f is a function of r^2 and η is a function of x, y, z.

In order for the new fields to be a solution of Maxwell equations in a vacuum (see the previous section), it is necessary for Equation (23) to satisfy the equations:

$$\nabla \cdot \mathbf{B}_0 = 0, \quad \nabla \cdot \mathbf{E}_0 = 0, \quad (24)$$

from which, given that $\nabla \cdot \mathbf{B}_{H,0} = 0$ and $\nabla \cdot \mathbf{E}_{H,0} = 0$, we get:

$$\frac{\nabla f}{f} \cdot \mathbf{B}_0 + \nabla\eta \cdot \mathbf{E}_0 = 0,$$
$$\frac{\nabla f}{f} \cdot \mathbf{E}_0 - \nabla\eta \cdot \mathbf{B}_0 = 0. \quad (25)$$

Using Equations (19), (21) and (23), Equation (25) can be written as:

$$\frac{\nabla f}{f} \cdot \mathbf{e}_1 + \nabla \eta \cdot \mathbf{e}_2 = 0,$$
$$\frac{\nabla f}{f} \cdot \mathbf{e}_2 - \nabla \eta \cdot \mathbf{e}_1 = 0. \tag{26}$$

Let us define $\gamma = r^2$. Since $f = f(r^2) = f(\gamma)$,

$$\frac{\nabla f}{f} = 2\Delta\,(x,y,z), \tag{27}$$

where we use the notation:

$$\Delta = \Delta(\gamma) = \frac{1}{f}\frac{df}{d\gamma}. \tag{28}$$

Taking into account Equation (21), Expression (26) leads to:

$$\nabla \eta \cdot \mathbf{e}_1 = x(r^2+1)\,\Delta,$$
$$\nabla \eta \cdot \mathbf{e}_2 = z(r^2+1)\,\Delta. \tag{29}$$

Since $\mathbf{e}_1$, $\mathbf{e}_2$, $\mathbf{e}_3$ form one basis of three-dimensional vectors Equation (22), we can write, using Equation (29),

$$\begin{aligned}\nabla \eta &= \frac{\nabla \eta \cdot \mathbf{e}_1}{e_1^2}\mathbf{e}_1 + \frac{\nabla \eta \cdot \mathbf{e}_2}{e_2^2}\mathbf{e}_2 + \frac{\nabla \eta \cdot \mathbf{e}_3}{e_3^2}\mathbf{e}_3 \\ &= \frac{4\Delta}{r^2+1}\,(x\,\mathbf{e}_1 + z\,\mathbf{e}_2) + \frac{4\delta}{(r^2+1)^2}\,\mathbf{e}_3,\end{aligned} \tag{30}$$

where we have defined $\delta = \nabla \eta \cdot \mathbf{e}_3$. Using Equation (21),

$$x\,\mathbf{e}_1 + z\,\mathbf{e}_2 = \frac{r^2+1}{2}\,(-z,-1,x) - \mathbf{e}_3, \tag{31}$$

so that:

$$\nabla \eta = 2\Delta\,(-z,-1,x) + \Sigma\,\mathbf{e}_3, \tag{32}$$

where:

$$\Sigma = \frac{4\delta}{(r^2+1)^2} - \frac{4\Delta}{r^2+1}. \tag{33}$$

We apply now the curl and project to the $\mathbf{e}_3$ direction, i.e, we apply the operator $\mathbf{e}_3 \cdot \nabla \times$ to Expression Equation (32). With the help of Equations (21) and (22), we obtain after some manipulations:

$$\Sigma = 2\Delta'\,(r^2 - y^2) + 2\Delta\left(1 - 2\frac{y^2+1}{r^2+1}\right), \tag{34}$$

where $\Delta' = d\Delta/d\gamma$. This shows that Σ depends only on $\gamma = r^2$ and y^2, so that:

$$\nabla \Sigma = 2\Sigma'\,(x,y,z) + 2y\,\frac{\partial \Sigma}{\partial y^2}\,(0,1,0). \tag{35}$$

From the condition $\nabla \times \nabla \eta = 0$, using Equations (32), (34) and (35), we get the following conditions:

$$0 = (\Sigma' + 2\Delta')(z + xy) - \left(\frac{r^2+1}{2}\Sigma' + \Sigma\right)z + \frac{\partial \Sigma}{\partial y^2}y(x - yz),$$

$$0 = (\Sigma' + 2\Delta')\left(\frac{r^2-1}{2} - y^2\right) + \frac{r^2+1}{2}\Sigma' + \Sigma + 2\left(\frac{r^2+1}{2}\Delta' + \Delta\right), \quad (36)$$

$$0 = (\Sigma' + 2\Delta')(x - yz) - \left(\frac{r^2+1}{2}\Sigma' + \Sigma\right)x - \frac{\partial \Sigma}{\partial y^2}y(z + xy). \quad (37)$$

Expressions Equations (36) and (37) can be simplified, and using Equation (34) to compute $\partial \Sigma / \partial y^2$, the previous system can be written as:

$$0 = \left(\frac{r^2+1}{2}\Sigma' + \Sigma\right) - (\Sigma' + 2\Delta')\left(1 + y^2\right), \quad (38)$$

$$0 = \frac{r^2+1}{2}(\Sigma' + 2\Delta') + 2\left(\frac{r^2+1}{2}\Delta' + \Delta\right), \quad (39)$$

$$0 = \frac{r^2+1}{2}(\Sigma' + 2\Delta') - 2\left(\frac{r^2+1}{2}\Delta' + \Delta\right). \quad (40)$$

The solution of this system of equations is:

$$0 = \frac{r^2+1}{2}\Sigma' + \Sigma, \quad (41)$$

$$0 = \Sigma' + 2\Delta', \quad (42)$$

$$0 = \frac{r^2+1}{2}\Delta' + \Delta, \quad (43)$$

which, after integration, gives:

$$\Delta = \frac{2m}{(r^2+1)^2}, \quad (44)$$

$$\Sigma = \frac{-4m}{(r^2+1)^2}, \quad (45)$$

where m is an integration constant that can be any real number. Inserting these solutions into Equation (32), after integration, η is found to be:

$$\eta = -m\frac{2y}{r^2+1}, \quad (46)$$

and using Equations (44) and (27) to solve for f gives:

$$f = \exp\left(m\frac{r^2-1}{r^2+1}\right), \quad (47)$$

where we have chosen the constants of integration so that this particular value is obtained.

Consequently, we found a solution of the form given by Equation (23) so that, in the MKS system of units, recovering the original notation X, Y, Z, R for the dimensionless coordinates, we get:

$$\mathbf{B}_0(\mathbf{r}) = \exp\left(m\frac{R^2-1}{R^2+1}\right)\left(\cos\left(m\frac{2Y}{R^2+1}\right)\mathbf{B}_{H,0} - \frac{1}{c}\sin\left(m\frac{2Y}{R^2+1}\right)\mathbf{E}_{H,0}\right),$$

$$\mathbf{E}_0(\mathbf{r}) = \exp\left(m\frac{R^2-1}{R^2+1}\right)\left(c\sin\left(m\frac{2Y}{R^2+1}\right)\mathbf{B}_{H,0} + \cos\left(m\frac{2Y}{R^2+1}\right)\mathbf{E}_{H,0}\right), \quad (48)$$

being $\mathbf{B}_{H,0}$, $\mathbf{E}_{H,0}$ from Equation (19):

$$\mathbf{B}_{H,0}(\mathbf{r}) = \frac{8\sqrt{a}}{\pi L_0^2 (R^2+1)^3} \left(Y - XZ, -X - YZ, \frac{X^2+Y^2-Z^2-1}{2} \right),$$

$$\mathbf{E}_{H,0}(\mathbf{r}) = \frac{8c\sqrt{a}}{\pi L_0^2 (R^2+1)^3} \left(\frac{X^2-Y^2-Z^2+1}{2}, -Z+XY, Y+XZ \right).$$

In Figures 1 and 2, we represent the field lines at $t = 0$ for the hopfion ($m = 0$) and the transformed fields for $m = 1$ and $m = 2$. It looks as if the closedness property of the hopfion field lines is broken, although part of the torus structure seems to be still present.

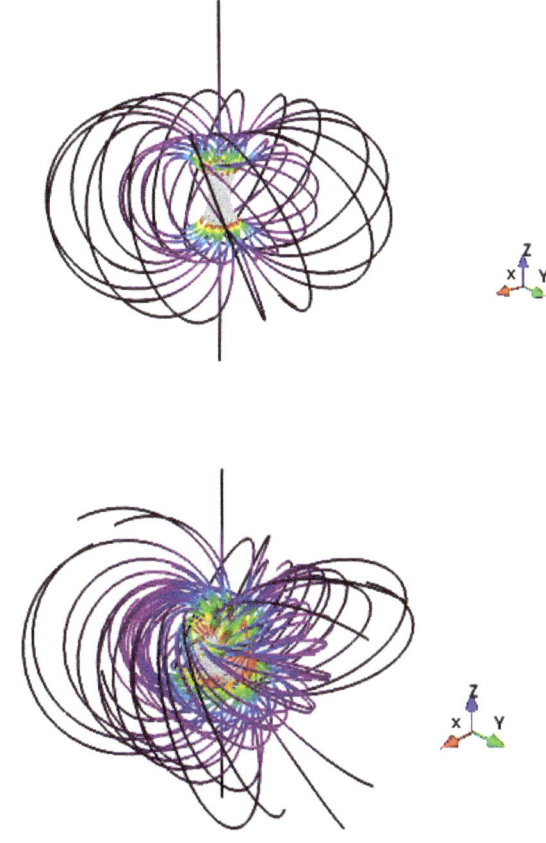

Figure 1. *Cont.*

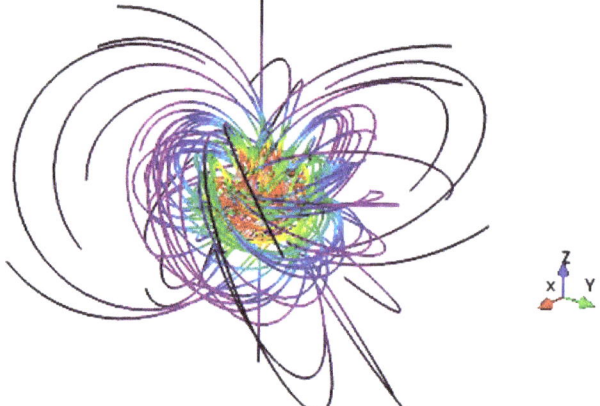

Figure 1. In the first figure (**top**), we represent some magnetic field lines for the initial value ($t = 0$) of the hopfion field, which corresponds to a value $m = 0$ in Equation (48). All the magnetic lines drawn are closed and linked to each other, which is the defining property of this field. In the second figure (**middle**), we plot some magnetic field lines for the transformed field with $m = 1$ in Equation (48), and in the third figure (**bottom**), some magnetic field lines for the transformed field with $m = 2$ in Equation (48) are drawn. The magnetic lines for the cases $m = 1$ and $m = 2$ seem to be unclosed in the numerics that give the lines plotted in the second and third figures.

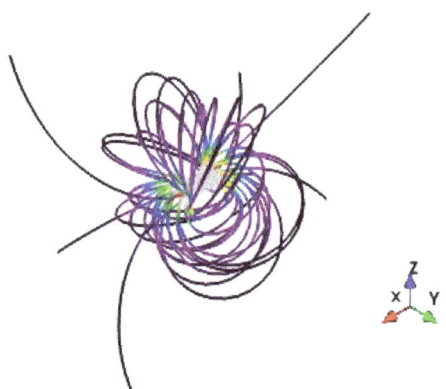

Figure 2. *Cont.*

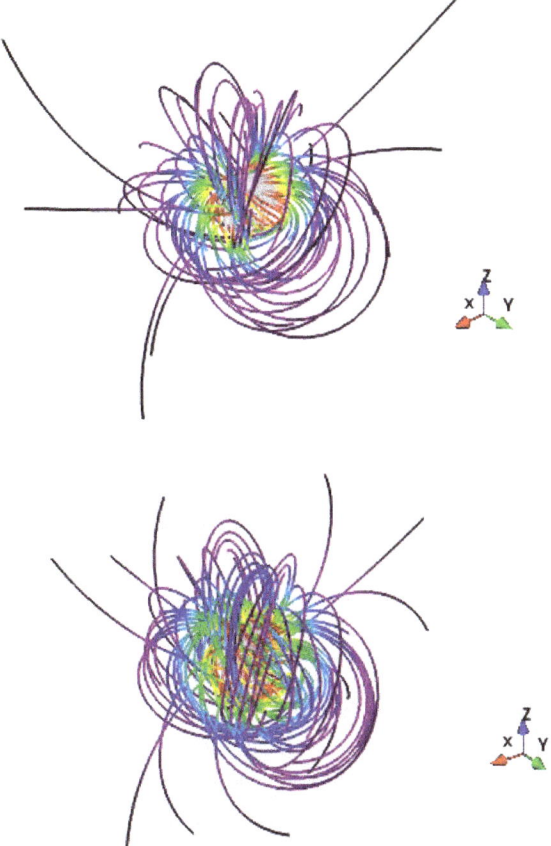

Figure 2. Same as Figure 1, but considering electric field lines. In the first figure (**top**), we represent electric field lines for the initial value ($t = 0$) of the hopfion field, all of which are closed and linked to each other. In the second figure (**middle**), we plot electric field lines for the transformed field with $m = 1$ in Equation (48), and in the third figure (**bottom**), we plot some electric field lines for the transformed field with $m = 2$ in Equation (48). The electric lines for the cases $m = 1$ and $m = 2$ seem to be unclosed in the numerics.

4. Bateman Formulation

After finding the fields at a particular time, we need to extend them to any time. To get the time-dependent expressions of the transformed fields Equation (48), we will make use of the Bateman formulation of null electromagnetic fields in a vacuum. In this section, we review very briefly Bateman's method.

In the case of Maxwell equations in a vacuum, it is useful to consider, instead of the magnetic and the electric fields separately, a complex combination of them called the Riemann–Silberstein vector **M** [18], which can be written as:

$$\mathbf{M} = \mathbf{E} + ic\,\mathbf{B}, \tag{49}$$

where c appears due to the different units of the magnetic and the electric fields. In terms of **M**, Maxwell equations in a vacuum Equations (3)–(6) read:

$$\nabla \cdot \mathbf{M} = 0, \tag{50}$$

$$\nabla \times \mathbf{M} = \frac{i}{c}\frac{\partial \mathbf{M}}{\partial t}. \tag{51}$$

The following expressions hold for the Riemann–Silberstein vector using Equations (7) and (8),

$$\begin{aligned}\mathbf{M} &= \frac{\sqrt{ac}}{2\pi i}\left(\frac{\nabla\bar{\theta}\times\nabla\theta}{(1+\theta\bar{\theta})^2} - i\frac{\nabla\bar{\phi}\times\nabla\phi}{(1+\phi\bar{\phi})^2}\right) \\ &= \frac{\sqrt{a}}{2\pi i}\left(\frac{\partial_t\bar{\phi}\nabla\phi - \partial_t\phi\nabla\bar{\phi}}{(1+\phi\bar{\phi})^2} + i\frac{\partial_t\bar{\theta}\nabla\theta - \partial_t\theta\nabla\bar{\theta}}{(1+\theta\bar{\theta})^2}\right).\end{aligned} \tag{52}$$

The formulation of electromagnetic fields constructed by Bateman in 1915 [19] can be used to study all null solutions of Maxwell equations in a vacuum [20]. The basic fields in this formulation are two complex functions $\alpha(\mathbf{r},t)$ and $\beta(\mathbf{r},t)$ so that the Riemann–Silberstein vector **M** of the electromagnetic field is written as:

$$\mathbf{M} = \mathbf{E} + ic\,\mathbf{B} = \frac{\sqrt{ac}}{\pi}\nabla\alpha\times\nabla\beta, \tag{53}$$

where the factor $\sqrt{ac}/\pi$ is chosen so that the comparison with the same vector Equation (52) is more direct. As a consequence, α and β are dimensionless functions of space and time.

Maxwell equations in a vacuum Equations (50) and (51) are satisfied by the electromagnetic field Equation (53) provided **M** can be also written as:

$$\mathbf{M} = \frac{i\sqrt{a}}{\pi}\left(\frac{\partial\alpha}{\partial t}\nabla\beta - \frac{\partial\beta}{\partial t}\nabla\alpha\right). \tag{54}$$

Equation $\nabla\cdot\mathbf{M}=0$ is satisfied by using Equation (53). To get the Maxwell equation $\nabla\times\mathbf{M}=i/c\,\partial\mathbf{M}/\partial t$, one can use Equation (54) in the left-hand side and Equation (53) in the right-hand side. The problem of finding solutions of Maxwell equations in a vacuum in the Bateman formulation is then reduced to solving the complex equation:

$$\nabla\alpha\times\nabla\beta = \frac{i}{c}\left(\frac{\partial\alpha}{\partial t}\nabla\beta - \frac{\partial\beta}{\partial t}\nabla\alpha\right), \tag{55}$$

for the complex fields $\alpha(\mathbf{r},t)$ and $\beta(\mathbf{r},t)$. A property of the electromagnetic fields in a vacuum constructed using the Bateman formulation is that they are null. Multiplying Equations (53) and (54), it is easily seen that $M^2=0$, so **M** defines a null electromagnetic field in a vacuum ($\mathbf{E}\cdot\mathbf{B}=0$ and $E^2-c^2B^2=0$).

We now obtain the hopfion in the Bateman formulation Equations (53) and (54). This was first obtained by Besieris and Shaarawi [21] and later by Kedia et al. [22] and Hoyos et al. [23], among others. Our results will be completely equivalent to the ones obtained in those references, although slightly different, since we are going to use Equations (11) and (12). We begin by writing the Riemann–Silberstein vector of the hopfion, using Equations (11) and (12), as:

$$\begin{aligned}\mathbf{M}_H &= \mathbf{E}_H + ic\,\mathbf{B}_H \\ &= \frac{\sqrt{ac}}{\pi}\nabla(u_2+iu_4)\times\nabla(u_3+iu_1) \\ &= \frac{i\sqrt{a}}{\pi}\left[\frac{\partial(u_2+iu_4)}{\partial t}\nabla(u_3+iu_1) - \frac{\partial(u_3+iu_1)}{\partial t}\nabla(u_2+iu_4)\right].\end{aligned}$$

Trivially, this is written in the Bateman formulation Equations (53) and (54) by identifying:

$$\begin{aligned} \alpha_H &= u_2 + i\, u_4, \\ \beta_H &= u_3 + i\, u_1. \end{aligned} \tag{56}$$

Making use of the values of the real scalar fields u_i Equations (13) and (16), we get:

$$\begin{aligned} \alpha_H &= \frac{Y + i(A-1)}{A + iT}, \\ \beta_H &= \frac{Z + iX}{A + iT}. \end{aligned} \tag{57}$$

Note that the results Equation (57) coincide with the ones found in [22] with changes $Z \to -Y$ and $Y \to -Z$, due to a different labeling of the axes.

An interesting observation about the Bateman formulation that we will use in this work is the following: every electromagnetic field $\mathbf{M}'$ constructed from a solution $\mathbf{M}(\alpha, \beta)$ of Maxwell's equations in a vacuum Equations (50) and (51) as:

$$\mathbf{M}' = g(\alpha, \beta)\, \mathbf{M}(\alpha, \beta), \tag{58}$$

where $g(\alpha, \beta)$ is an arbitrary function of the complex fields α and β is also a solution [19].

This property was exploited by Kedia et al. in [22] to find a set of null solutions of the Maxwell equations in a vacuum that generalize the Hopfion and give field lines that are linked torus knots at $t = 0$. The solutions they found can be written in the Bateman formulation Equations (53) and (54) using two positive integer numbers m and n as:

$$\begin{aligned} \alpha_m &= (u_2 + i\, u_4)^m = \left(\frac{Y + i(A-1)}{A + iT}\right)^m, \\ \beta_n &= (u_3 + i\, u_1)^n = \left(\frac{Z + iX}{A + iT}\right)^n, \end{aligned} \tag{59}$$

Taking again for the u_i's the hopfion values Equations (13) and (16), we can write Equation (59) as:

$$\begin{aligned} \alpha_m &= (\alpha_H)^m, \\ \beta_n &= (\beta_H)^n, \end{aligned} \tag{60}$$

Using Equations (53) and (54), we get for this case:

$$\begin{aligned} \mathbf{M}_{nm} &= \frac{\sqrt{ac}}{\pi}\, \nabla \alpha_m \times \nabla \beta_n = \frac{i\sqrt{a}}{\pi} \left(\frac{\partial \alpha_m}{\partial t} \nabla \beta_n - \frac{\partial \beta_n}{\partial t} \nabla \alpha_m\right) \\ &= g(\alpha_H, \beta_H)\, \mathbf{M}_H, \end{aligned} \tag{61}$$

with:

$$g(\alpha_H, \beta_H) = (m\, n)\, \alpha_H^{m-1} \beta_H^{n-1}. \tag{62}$$

Further generalizations for generating other fields can be found in [23].

5. A New Set of Null Electromagnetic Fields

In this section, we extend the transformations for the fields given by Equation (48) at a particular time to any time exploiting the properties of the Bateman formulation we have just reviewed in the previous section. Thus, we will get the new set of null electromagnetic fields.

As pointed out earlier, the Riemann–Silberstein vector allows writing the transformed fields at a particular time in a very compact form, which sheds some light on the transform valid for any time. For the transformed fields at $t = 0$ Equation (48), we get:

$$\mathbf{M}_0 = \exp\left(-im\frac{2Y + i(R^2 - 1)}{R^2 + 1}\right) \mathbf{M}_{H,0}. \tag{63}$$

$\mathbf{M}_{H,0}$ being the Riemann–Silberstein vector of the hopfion at $t = 0$, i.e.,:

$$\mathbf{M}_{H,0} = \frac{8c\sqrt{a}}{\pi L_0^2(1+R^2)^3}\left(\frac{X^2 - Y^2 - Z^2 + 1}{2} + i(Y - XZ),\right.$$
$$\left.(XY - Z) - i(X + YZ), (Y + XZ) + i\frac{X^2 + Y^2 - Z^2 - 1}{2}\right). \tag{64}$$

The key point is to observe that we can write Equation (63) as:

$$\mathbf{M}_0 = \exp\left[-im\left(u_{2,0} + i\,u_{4,0}\right)\right] \mathbf{M}_{H,0}, \tag{65}$$

where, according to Equations (14) and (16), $u_{2,0}$ and $u_{4,0}$ are, respectively, the values of u_2 and u_4 at $t = 0$.

Invoking the property Equation (58), we can extend the transformed fields for all time:

$$\mathbf{M} = \exp\left[-im\left(u_2 + i\,u_4\right)\right] \mathbf{M}_H, \tag{66}$$

where u_2 and u_4 are given by Equations (14) and (16) and $\mathbf{M}_H$ is the Riemann–Silberstein vector of the Hopfion Equation (56). This means that we can express $\mathbf{M}$ as:

$$\mathbf{M} = \frac{c\sqrt{a}}{\pi} \nabla\left(\frac{i}{m}e^{-im(u_2+i u_4)}\right) \times \nabla\left(u_3 + i\,u_1\right) \tag{67}$$

$$= \frac{i\sqrt{a}}{\pi}\left[\frac{\partial\left(\frac{i}{m}e^{-im(u_2+i u_4)}\right)}{\partial t}\nabla\left(u_3 + i\,u_1\right) - \frac{\partial\left(u_3 + i\,u_1\right)}{\partial t}\nabla\left(\frac{i}{m}e^{-im(u_2+i u_4)}\right)\right]. \tag{68}$$

Moreover, we can generalize this solution by playing the same game with the complex field $u_3 + i\,u_1$. We arrive then at our final expression for a new set of electromagnetic fields constructed from the hopfion by dilation and rotation. These fields are given by the Riemann–Silberstein vector defined as:

$$\mathbf{M}_{DR} = \frac{c\sqrt{a}}{\pi}\nabla\left(\frac{i}{m}e^{-im(u_2+i u_4)}\right) \times \nabla\left(\frac{i}{n}e^{-in(u_3+i u_1)}\right) \tag{69}$$

$$= \frac{i\sqrt{a}}{\pi}\left[\frac{\partial\left(\frac{i}{m}e^{-im(u_2+i u_4)}\right)}{\partial t}\nabla\left(\frac{i}{n}e^{-in(u_3+i u_1)}\right) - \frac{\partial\left(\frac{i}{n}e^{-in(u_3+i u_1)}\right)}{\partial t}\nabla\left(\frac{i}{m}e^{-im(u_2+i u_4)}\right)\right].$$

that, by construction, are null electromagnetic fields and satisfy Maxwell equations in a vacuum when u_1, u_2, u_3, u_4 are the ones defined in Expressions (13)–(16). In a more compact notation,

$$\mathbf{M}_{DR} = e^{-i[n(u_3+i u_1)+m(u_2+i u_4)]} \mathbf{M}_H. \tag{70}$$

This expression shows what kind of new solutions have been found. They correspond to a dilation of the magnetic and electric fields of the Hopfion plus a rotation in the plane $(\mathbf{E}_H, \mathbf{B}_H)$. Both the dilation and the rotation parameters depend on the u_i's variables and on the two constants m and n, which can be seen by writing the new solutions as:

$$\begin{aligned}
\mathbf{B}_{DR}(\mathbf{r},t) &= e^{(n u_1 + m u_4)} \left(\cos(n u_3 + m u_2) \, \mathbf{B}_H(\mathbf{r},t) - \frac{1}{c} \sin(n u_3 + m u_2) \, \mathbf{E}_H(\mathbf{r},t) \right), \\
\mathbf{E}_{DR}(\mathbf{r},t) &= e^{(n u_1 + m u_4)} \left(c \sin(n u_3 + m u_2) \, \mathbf{B}_H(\mathbf{r},t) + \cos(n u_3 + m u_2) \, \mathbf{E}_H(\mathbf{r},t) \right).
\end{aligned} \quad (71)$$

In Figures 3 and 4, we represent the field lines for the transformed fields with $(n,m) = (1,2)$ at $t = 0$. As before, the closedness property seems to disappear.

Figure 3. In this figure, we represent some magnetic field lines for the initial value ($t = 0$) of the magnetic field given in Equation (71) with values $n = 1$ and $m = 2$. As in the cases shown in the second and third plots of Figure 1, these magnetic lines seem to be unclosed in the numerics.

Figure 4. Same as Figure 3, but considering electric field lines. We plot some electric field lines for the initial value ($t = 0$) of the electric field given in Equation (71) with values $n = 1$ and $m = 2$. As in the cases shown in the second and third plots of Figure 2, these electric lines seem to be unclosed in the numerics.

6. Four-Potentials and Helicities

The Bateman formalism is also to find the electromagnetic four-potentials for the solutions that can be written as in Equations (53) and (54). In this section, we will give the recipe and calculate them for the new solutions obtained in the previous section.

It is standard to define the electromagnetic four-potential $A^\mu = (A^0, \mathbf{A})$ in Minkowski spacetime, with $\mu = 0,1,2,3$ and a metric given by diagonal elements $(1,-1,-1,-1)$, so that $\mathbf{B} = \nabla \times \mathbf{A}$, $\mathbf{E} = -c\left(\nabla A^0 + (1/c)\partial \mathbf{A}/\partial t\right)$. It is possible only in a vacuum to define (see [1] for example) a second electromagnetic potential $C^\mu = (C^0, \mathbf{C})$ that satisfies $\mathbf{E} = \nabla \times \mathbf{C}$, $\mathbf{B} = (1/c)\left(\nabla C^0 + (1/c)\partial \mathbf{C}/\partial t\right)$. Let us consider the complex combination:

$$N^\mu = (N^0, \mathbf{N}) = C^\mu + ic\, A^\mu = (C^0 + ic\, A^0, \mathbf{C} + ic\, \mathbf{A}). \tag{72}$$

Then, the Riemann–Silberstein vector satisfies:

$$\mathbf{M} = \nabla \times \mathbf{N} = i\left(\nabla N^0 + \frac{1}{c}\frac{\partial \mathbf{N}}{\partial t}\right). \tag{73}$$

Now, consider the situation in which the electromagnetic field can be written in the Bateman formalism as:

$$\mathbf{M} = \frac{\sqrt{ac}}{\pi} \nabla\alpha \times \nabla\beta = \frac{i\sqrt{a}}{\pi}\left(\frac{\partial \alpha}{\partial t}\nabla\beta - \frac{\partial \beta}{\partial t}\nabla\alpha\right). \tag{74}$$

From Equation (73), it can be seen that an electromagnetic potential for Equation (74) is:

$$N^0 = -\frac{\sqrt{a}}{2\pi}\left(\alpha\frac{\partial \beta}{\partial t} - \beta\frac{\partial \alpha}{\partial t}\right),$$

$$\mathbf{N} = \frac{c\sqrt{a}}{2\pi}(\alpha\nabla\beta - \beta\nabla\alpha). \tag{75}$$

We have also a gauge degree of freedom to choose any other electromagnetic potential $\tilde{N}^\mu$ from Equation (75) as:

$$\tilde{N}^\mu = N^\mu + \partial^\mu f(\mathbf{r},t). \tag{76}$$

The magnetic h_m and electric h_e helicities [24] of an electromagnetic field in a vacuum can be defined, in angular momentum units, as:

$$h_m = \frac{\varepsilon_0 c}{2}\int \mathbf{A}\cdot\mathbf{B}\,d^3r,$$

$$h_e = \frac{\varepsilon_0}{2c}\int \mathbf{C}\cdot\mathbf{E}\,d^3r, \tag{77}$$

where ε_0 is the electric permittivity of a vacuum. The magnetic helicity measures the mean value of the Gauss linking number of the magnetic field lines [25], and the electric helicity plays the same role for the electric field. For null electromagnetic fields in a vacuum, $h_e = h_m$, and moreover, both helicities remain independent of time [2]. For an electromagnetic field expressed in the Bateman formalism as Equation (74) with the electromagnetic potential given by Equation (75), it is trivial to check that:

$$h_e - h_m = \mathrm{Re}\left(\frac{\varepsilon_0}{2c}\int \mathbf{N}\cdot\mathbf{M}\,d^3r\right) = 0, \tag{78}$$

showing that we are in the case of null fields.

The sum of both helicities is the electromagnetic helicity [26],

$$h_{em} = h_e + h_m = \mathrm{Re}\left(\frac{\varepsilon_0}{2c}\int \tilde{\mathbf{N}}\cdot\mathbf{M}\,d^3r\right), \tag{79}$$

and it can be written in the Bateman formalism using again Equations (74) and (75) as:

$$h_{em} = \frac{\varepsilon_0 ac}{4\pi^2} \operatorname{Re}\left[\int (\bar{\alpha}\nabla\alpha \cdot (\nabla\beta \times \nabla\bar{\beta})\right.$$
$$\left. + \bar{\beta}\nabla\beta \cdot (\nabla\alpha \times \nabla\bar{\alpha})\right] d^3r. \tag{80}$$

with a bar over a letter meaning complex conjugation.

In particular, for the hopfion Equation (56), we have, due to Equation (57), for the potential Equation (75),

$$N_H^0 = -\frac{i\sqrt{a}}{2\pi} \frac{T(Z+iX)}{(A+iT)^2},$$
$$\mathbf{N}_H = \frac{c\sqrt{a}}{2\pi(A+iT)^2}\left[1 - A + X^2 + i(Y - XZ),\right. \tag{81}$$
$$\left. -Z + XY - i(X+YZ), Y + XZ + i(A - 1 - Z^2)\right],$$

results that are, except a gauge transformation, similar to the ones found in [27]. The electromagnetic helicity of the hopfion, according to Equation (80), results in:

$$h_{em,H} = \varepsilon_0 ac, \tag{82}$$

which can be taken as the unit of electromagnetic helicity for the hopfion and related electromagnetic fields in a vacuum [24].

Let us consider now the electromagnetic potential of the solutions Equation (71) found in the previous section. For them,

$$\alpha_{DR} = (i/m)\exp(-im(u_2 + iu_4)),$$
$$\beta_{DR} = (i/n)\exp(-in(u_3 + iu_1)). \tag{83}$$

and from Equation (75), the potentials read:

$$N_{DR}^0 = \frac{i\sqrt{a}}{2\pi mn} e^{-i[n(u_3+iu_1)+m(u_2+iu_4)]} \frac{\partial}{\partial t}[m(u_2+iu_4) - n(u_3+iu_1)],$$
$$\mathbf{N}_{DR} = \frac{-ic\sqrt{a}}{2\pi mn} e^{-i[n(u_3+iu_1)+m(u_2+iu_4)]} \nabla[m(u_2+iu_4) - n(u_3+iu_1)], \tag{84}$$

where the u_i's are given in Equations (13)–(16). The magnetic and electric helicities of these solutions have the same value, since the field is null. The electromagnetic helicity is, according to Equations (80) and (82):

$$h_{em} = \frac{h_{em,H}}{2\pi^2}\int \frac{e^{2(nu_1+mu_4)}}{mn}(m\nabla u_4 - n\nabla u_1)\cdot(\nabla u_2 \times \nabla u_3)\, d^3r.$$

7. Conclusions

In this work, we studied symmetry transformations at a particular time for generating new solutions of Maxwell equations in a vacuum. The transformations need to satisfy some constraints. We then made use of the Bateman formulation to get new null electromagnetic fields at any time.

Another important result presented was the method of using the Bateman formulation to find the electromagnetic potentials for null solutions of Maxwell equations in a vacuum. This method allowed us to compute their magnetic and electric helicities.

We found numerically that the topology seemed to not be conserved, and the closedness of the electric and magnetic field lines was broken for the new fields. This is a fact that has been observed in other solutions that used the hopfion and the Bateman construction for generating other fields [22]. However, we cannot exclude that it was due to numerical accuracy. This point together with the linking of the field lines deserves future investigation, and it might shed some light on the complex dynamics of these knotted fields and how to generate them in the laboratory.

The research on electromagnetic knots belongs to the research of the classical theory of particles and fields, but the techniques and concepts involved might be extended to other areas, such as dynamical systems, liquid crystals, stellar plasma configurations, topological insulators, etc. Many of the consequences remain to be explored.

Author Contributions: All the authors contributed equally to all the parts of this work.

Funding: This research was funded by the Spanish Ministry of Economy and Competitiveness grant number ESP2017-86263-C4-3-R.

Conflicts of Interest: The authors declare no conflict of interest.

References

1. Arrayás, M.; Bouwmeester, D.; Trueba, J.L. Knots in electromagnetism. *Phys. Rep.* **2017**, *667*, 1–61. [CrossRef]
2. Rañada, A.F.; Trueba, J.L. Two properties of electromagnetic knots. *Phys. Lett. A* **1997**, *232*, 25–33. [CrossRef]
3. Rañada, A.F.; Trueba, J.L. Topological Electromagnetism with Hidden Nonlinearity. In *Modern Nonlinear Optics*, 3rd ed.; Evans, M.W., Ed.; John Wiley & Sons: New York, NY, USA, 2001; pp. 197–253.
4. Irvine, W.T.; Bouwmeester, D. Linked and knotted beams of light. *Nat. Phys.* **2008**, *4*, 716–720. [CrossRef]
5. Robinson, I. Null electromagnetic fields. *J. Math. Phys.* **1961**, *2*, 290. [CrossRef]
6. Rañada, A.F. A topological theory of the electromagnetic field. *Lett. Math. Phys.* **1989**, *18*, 97–106. [CrossRef]
7. Rañada, A.F. Knotted solutions of the Maxwell equations in vacuum. *J. Phys. A Math. Gen.* **1990**, *23*, L815. [CrossRef]
8. Rañada, A.F. Topological electromagnetism. *J. Phys. A Math. Gen.* **1992**, *25*, 1621. [CrossRef]
9. Rañada, A.F.; Trueba, J.L. Electromagnetic knots. *Phys. Lett. A* **1995**, *202*, 337–342. [CrossRef]
10. Irvine, W.T.M. Linked and knotted beams of light, conservation of helicity and the flow of null electromagnetic fields. *J. Phys. A Math. Theor.* **2010**, *43*, 385203. [CrossRef]
11. Dalhuisen, J.W.; Bouwmeester, D. Twistors and electromagnetic knots. *J. Phys. A Math. Theor.* **2012**, *45*, 135201. [CrossRef]
12. Arrayás, M.; Trueba, J.L. Exchange of helicity in a knotted electromagnetic field. *Ann. Phys.* **2012**, *524*, 71–75. [CrossRef]
13. Enk, S.J. Covariant description of electric and magnetic field lines of null fields: Application to Hopf-Rañada solutions. *J. Phys. A Math. Theor.* **2013**, *46*, 175204. [CrossRef]
14. Arrayás, M.; Trueba, J.L. A class of non-null toroidal electromagnetic fields and its relation to the model of electromagnetic knots. *J. Phys. A Math. Theor.* **2015**, *48*, 025203. [CrossRef]
15. Arrayás, M.; Trueba, J.L. Collision of two hopfions. *J. Phys. A Math. Theor.* **2017**, *50*, 085203. [CrossRef]
16. Arrayás, M.; Trueba, J.L. On the fibration defined by the field lines of a knotted class of electromagnetic fields at a particular time. *Symmetry* **2017**, *9*, 218. [CrossRef]
17. Arrayás, M.; Trueba, J.L. Motion of charged particles in a knotted electromagnetic field. *J. Phys. A* **2010**, *43*, 235401. [CrossRef]
18. Bialynicki-Birula, I.; Bialynicka-Birula, Z. The role of the Riemann-Silberstein vector in Classical and Quantum Theories of Electromagnetism. *J. Phys. A Math. Theor.* **2013**, *46*, 053001. [CrossRef]
19. Bateman, H. *The Mathematical Analysis of Electrical and Optical Wave-Motion*; Dover: New York, NY, USA, 1915.
20. Hogan, P. Bateman electromagnetic waves. *Proc. R. Soc. A* **1984**, *396*, 199. [CrossRef]
21. Besieris, I.M.; Shaarawi, A. Hopf-Rañada linked and knotted light beam solution viewed as a null field. *Opt. Lett.* **2009**, *34*, 3887–3889. [CrossRef]
22. Kedia, H.; Bialinicki-Birula, I.; Peralta-Salas, D.; Irvine, W.T.M. Tying knots in beams of light. *Phys. Rev. Lett.* **2013**, *111*, 150404. [CrossRef]

23. Hoyos, C.; Sircar, N.; Sonnenschein, J. New knotted solutions of Maxwell's equations. *J. Phys. A* **2015**, *48*, 255204. [CrossRef]
24. Arrayás, M.; Trueba, J.L. Spin-Orbital momentum decomposition and helicity exchange in a set of non-null knotted electromagnetic fields. *Symmetry* **2018**, *10*, 88. [CrossRef]
25. Moffatt, H.K.; Ricca, R.L. Helicity and the Calugareanu Invariant. *Proc. R. Soc. Lond. A* **1992**, *439*, 411–429. [CrossRef]
26. Trueba, J.L.; Rañada, A.F. The electromagnetic helicity. *Eur. J. Phys.* **1996**, *17*, 141. [CrossRef]
27. Rañada, A.F.; Tiemblo, A.; Trueba, J.L. Time evolving potentials for electromagnetic knots. *Int. J. Geom. Methods Mod. Phys.* **2017**, *14*, 1750073. [CrossRef]

© 2019 by the authors. Licensee MDPI, Basel, Switzerland. This article is an open access article distributed under the terms and conditions of the Creative Commons Attribution (CC BY) license (http://creativecommons.org/licenses/by/4.0/).

Article

On the Computation of the Dispersion Diagram of Symmetric One-Dimensionally Periodic Structures

Francisco Mesa [1,*,†], **Raúl Rodríguez-Berral** [1,†] **and Francisco Medina** [2,†]

1 Department of Applied Physics 1, ETS de Ingeniería Informática, Universidad de Sevilla,
  Av. Reina Mercedes s/n, 41012 Sevilla, Spain; rrberral@us.es
2 Department of Electronics & Electromagetism; Facultad de Física, Universidad de Sevilla,
  Av. Reina Mercedes s/n, 41012 Sevilla, Spain; medina@us.es
* Correspondence: mesa@us.es; Tel.: +34-954-55-6155
† These authors contributed equally to this work.

Received: 13 July 2018; Accepted: 27 July 2018; Published: 1 August 2018

Abstract: A critical discussion on the computation of the dispersion diagram of electromagnetic guiding/radiating structures with one-dimensional periodicity using general-purpose electromagnetic simulation software is presented in this work. In these methods, full-wave simulations of finite sections of the periodic structure are combined with appropriate simplifying network models. In particular, we analyze the advantages and limitations of two different combined methods, with emphasis on the determination of their range of validity. Our discussion is complemented with several selected numerical examples in order to show the most relevant aspects that a potential user of these methods should be aware of. Special attention is paid to the relevant role played by the high-order coupling between adjacent unit cells and between the two halves of unit cells exhibiting reflection, inversion, and glide symmetries.

Keywords: periodic structures; dispersion diagram; high-order coupling; glide symmetry

1. Introduction

Many practical microwave/antenna devices find their fundamental operating mechanisms in the behavior of electromagnetic waves in a periodic environment [1–4]. Examples of this are waveguide/printed-line periodic filters [2,5], metamaterial-inspired transmission lines [6,7], periodic leaky-wave antennas [8], frequency selective surfaces (FSS) [9], reflect/transmit-arrays [10,11], metasurfaces [12], etc. In all these problems, many of the relevant transmission, reflection, and/or absorption characteristics of the periodic (or quasi-periodic) finite device can be explained from the knowledge of the dispersion diagram of the corresponding infinitely periodic structure. As is well known, the treatment of these structures can be reduced to deal only with the unit cell of the periodic structure. Thus, in every of the above mentioned problems we can identify a basic propagation and/or radiation problem involving discontinuities in a generalized waveguiding system subject to periodic boundary conditions. The waveguiding system can be a standard metallic waveguide [2], a generalized waveguide [13,14] (as the one typically found in the treatment of FSSs [15,16]), printed lines [7,17], substrate integrated waveguides [18–20], etc. The periodic boundary conditions can appear either in the walls of the waveguiding system (transverse periodicity) and/or along the propagation direction (longitudinal periodicity). If none of the boundaries of the waveguiding system is open to free space, the periodic electromagnetic wave problem can be solved by means of a Floquet analysis of the structure [21,22] that involves only a discrete spectrum [23]. If there are open boundaries in the waveguiding system, the continuous spectrum should also be taken into account by means of its necessary integral representation [3,23,24]. In any case, the dispersion diagram of the periodic structure can be obtained after solving the non-linear eigenvalue problem that results from the application of Maxwell's equations with the appropriate boundary conditions to the considered unit cell [3,23].

Likewise other non-linear electromagnetic eigenvalue problems, the obtaining of the eigenvalues requires the searching for complex zeros of a given determinantal equation. In the present problem, the eigenvalues correspond to the wavenumbers of the propagating modes, say $k_z = \beta - j\alpha$, with β being the phase constant and α the attenuation constant (α accounts either for the evanescent/complex nature of the mode or for the presence of material and/or radiation losses). As well reported in the literature, the zero-searching task in the complex plane is not trivial at all because of its intrinsic difficulty to be systematized into a general algorithm that can easily be implemented in an unattended computer code [25–27]. Also, to the best of the authors' knowledge, commercial electromagnetic simulators (for instance [28,29]) provide systematically the frequency behavior of only the real part of the complex wavenumber (an eigenvalue problem is defined by imposing a given phase shift between the boundaries of the unit cell, and the corresponding eigen-frequency is then computed).

As an alternative to solving the above non-linear eigenvalue problem with its intrinsic cumbersome task of searching for zeros in the complex plane, different procedures that make use of general-purpose electromagnetic simulators (or even measurements) have been reported in the literature. One of these procedures involves the analysis of one or two sections of the periodic structure involving several unit cells in order to extract the dispersion relation from the different values of the associated transmission matrices [18,19,30–32]. Implicit in the above method is the modeling of the periodic structure as a cascade of identical two-port (or multi-port) equivalent networks [3,4]. The decomposition in two-port equivalent networks is valid when the interaction between adjacent cells is well accounted for by only the fundamental mode of the waveguiding structure. If higher order modes and/or the continuous spectrum take part in this interaction, then multi-port equivalent networks are necessary [3,33]. In many published works on this topic, the application of the above procedures requires the implementation of in-house computer codes [34–36]. Usually these codes are not easy to be reproduced by (or distributed to) other authors and also hard to be generalized to cases other than the particular ones treated in the corresponding papers. Certainly, wide distribution and versatility of the software tool are two well-known and very relevant characteristics of commercial electromagnetic simulators. In consequence, the development of combined approaches that can take advantage of these features of commercial simulators and can be complemented with a simple in-house post-processing stage is becoming more and more convenient [37–40]. Some more-simplified approaches involving the simulation of a single unit cell (or even half unit cell) have also been proposed [41]. Thus, the main goal of the present work will be to go over some of these approaches in order to discuss what is the optimized hybrid method that, making use of commercial software, can efficiently provide the dispersion relation of periodic structures. The presence of internal symmetries in the unit cell will be of special interest in the frame of this research.

The paper is organized as follows. Section 2 gives an overview of two of the most usual methods to deal with one-dimensionally periodic guiding structures and also discusses how to take advantage of the possible symmetries in the unit cell. Section 3 discusses the conclusions derived in the previous section in connection with the numerical results obtained for two different structures: printed periodic microstrip lines and paralle-plate waveguides with periodic corrugations. Finally, a summary of our main conclusions is reported in Section 4.

2. Methods of Analysis

As it has been mentioned above, there are basically two possible rigorous procedures to obtain the dispersion diagram of periodic structures [3,4]: (i) the solution of the corresponding eigenvalue problem associated with a unit cell subject to periodic boundary walls (PBW); and (ii) the extraction of the dispersion curves from the post-processing of the ABCD matrix of a cascade of multi-modal (multi-port) equivalent networks. This second approach is the one reported in the literature when the dispersion diagram of the periodic structure is computed from the full-wave simulation results of a macro-cell made up of several unit cells [18,19,37–39], and will be the subject of discussion of the present work. Certainly, the full-wave simulator employed to characterize the macro-cell does take into account all the possible interactions between adjacent cells (which justifies its identification with a cascade of multi-port equivalent networks). This method has several drawbacks such as the intense computational load required to analyze a macro-cell involving many unit cells (as required in many practical problems) [39,41], the appearance of spurious solutions, and the ambiguity of the phase constant outside the first fictitious Brillouin zone imposed by the repetition of the unit cell [39]. These inherent difficulties have motivated the search for approaches that can overcome them, as for example those reported in [39,41]. The procedure given in [39] can solve the drawback related to the appearance of spurious solutions at the expense of increasing the computational load. However, the procedure reported in [41] apparently overcomes all the mentioned drawbacks given that it only involves full-wave simulations of a single unit cell (or even half this unit cell under appropriate symmetry conditions) bounded by electric and/or magnetic walls along the direction of periodicity. Unfortunately, the authors of the present work have not been able to reproduce the expected good results of [41] in their own research and have found some reasons to justify this fact. The theoretical and numerical results of the authors in this research are discussed below.

An example of a possible periodic configuration of interest within the frame of the present work is schematically shown in Figure 1a. This arrangement consists of a section of three unit cells of a longitudinally (along z-direction) periodic structure. The figure could represent either the longitudinal cut of a rectangular or parallel-plate waveguide (PPW) with vertical (E-plane) stubs (in such case the solid color represents the interior of the metallic waveguide) or the top view of the layout corresponding to a printed microstrip line periodically loaded with stubs at the right and left sides. In any case, the structure can be modeled by a cascade of equivalent networks as shown in Figure 1b,c. Panel (b) corresponds to a cascade of two-port networks and panel (c) to a cascade of multi-port networks. In Figure 1c it is assumed that port (1) is associated with the fundamental mode, port (2) with the first higher-order mode, and so on. As usual, the input and output ports of the whole structure are associated with the fundamental mode. In the example of Figure 1c, the interaction between adjacent cells is assumed to be accounted for by both the first and second modes, with the remaining modes being considered "localized" modes and therefore only contributing as lumped elements in the equivalent network [15,33,42]. Clearly, the network shown in Figure 1b is a simplification of the one in Figure 1c provided that the fundamental mode is the only relevant mode in the interaction between adjacent cells, as implicitly or explicitly assumed in many works in the literature [43]. Note that although, in principle, the boundaries of the unit cell along the z-direction can be arbitrarily defined (provided the period is respected), it is convenient to set such boundaries far apart from the discontinuities (stubs in this case), in a region where the uniform housing guiding system (rectangular waveguide, PPW or microstrip) is clearly recognizable. If the unit cell has internal symmetries, the boundaries should be chosen in such a way that these symmetries are preserved. Thus, for instance, the limits of the unit cell might have been chosen between the two stubs represented in Figure 1a, but this would not be a good choice if a model like the one in Figure 1b or Figure 1c is intended to be used.

Next, two commonly used methods reported in the literature to obtain the dispersion relation of waveguiding periodic structures along the longitudinal direction will be presented and critically discussed. Both methods combine full-wave simulations coming from commercial electromagnetic solvers with some post-processing to give the dispersion relation of the structures in a systematic way.

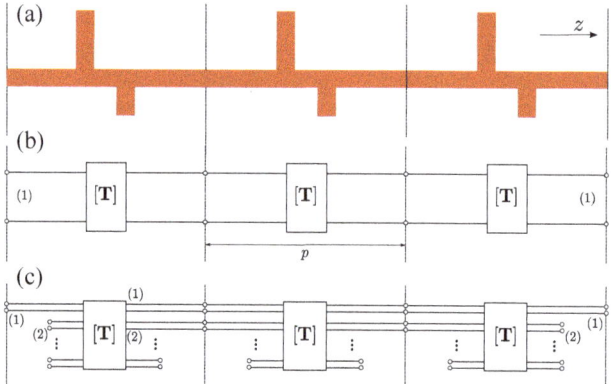

Figure 1. Periodic structure under study, period p. (**a**) Cascade of 1-port equivalent networks; (**b**) Cascade of multi-port equivalent networks; (**c**) Possible actual appearance of the physical periodic structure with just 3 unit cells.

2.1. Method A

For a generic periodic configuration as the one in Figure 1a, and assuming that an appropriate deembedding procedure has been implemented [18,19] to cancel out the undesirable effects caused by the practical feeding of the structure, a very general and efficient method proposed in the literature to obtain the dispersion diagram is based on the full-wave simulation of the finite N-cell structure to obtain, in a first step, the corresponding total transmission matrix associated with the input and output fundamental mode. Here it should again be noted that the boundaries of the unit cell of the structure should conveniently be chosen, if possible, so that most of the interactions between cells is carried out by means of the fundamental mode (the unit cell shown in Figure 1a is an example of this convenient choice). The transmission matrix, $[\mathbf{T}_N]$, corresponding to a cascade of N unit cells can formally be written as

$$[\mathbf{T}_N] = [\mathbf{T}]^N \tag{1}$$

where $[\mathbf{T}]$ stands for the unit-cell ABCD matrix corresponding to the fundamental mode in an scenario where the higher-order mode interaction between cells has been appropriately taken into account. It should be noted that, only under this assumption, the cascade of multi-port equivalent networks has formally been expressed as a cascade of "effective" two-port ABCD matrices (as in Figure 1b), which would be computed as

$$[\mathbf{T}] = \sqrt[N]{[\mathbf{T}_N]} = \begin{bmatrix} A_p & B_p \\ C_p & D_p \end{bmatrix} \tag{2}$$

where the subindex p indicates that the elements refer to a region of length p (that is, the period of the unit cell). The term "effective" comes along with this $[\mathbf{T}]$ matrix to point out that this matrix is not the standard ABCD matrix of an isolated unit cell interacting with adjacent cells only through the fundamental mode (indeed, the "effective" $[\mathbf{T}]$ matrix depends on the number of unit cells in the cascade). The dispersion relation of the periodic configuration would then be given by [3,4]

$$\cosh(\gamma p) = \frac{A_p + D_p}{2} \tag{3}$$

or by the spurious-free procedure given in [39]. In the above equation γ is the propagation constant, which is related to the wavenumber by $\gamma = jk_z = j\beta + \alpha$.

2.2. Method B

In the above discussion of the periodic generic configuration of Figure 1, it was implicitly assumed that the procedure reported in [41] could not be applied because of the lack of symmetry in the unit cell. However, if the unit cell does have symmetries as those shown in Figure 2b,c, then the authors in [41] propose to exploit such symmetries to express the dispersion relation in terms of the properties of just one half of the unit cell. In general, the smaller the structure to be analyzed with the full-wave simulator, the more accurate and spurious-free the computed numerical solution. Following [41] it will be assumed the existence of a cascade of two-port transmission matrices as in Figure 2a, where the symmetry of the unit cell is reflected by the following form of the matrices:

$$[\mathbf{T}'] = \begin{bmatrix} A_{p/2} & B_{p/2} \\ C_{p/2} & D_{p/2} \end{bmatrix} \quad , \quad [\mathbf{T}''] = \begin{bmatrix} D_{p/2} & B_{p/2} \\ C_{p/2} & A_{p/2} \end{bmatrix} \tag{4}$$

(subindex $p/2$ stands for the fact that only half the unit cell is considered for the definition of each of these auxiliary transfer matrices). The transmission matrix of the global unit cell is then given by

$$[\mathbf{T}] = [\mathbf{T}'][\mathbf{T}''] = \begin{bmatrix} A_{p/2}D_{p/2} + B_{p/2}C_{p/2} & 2A_{p/2}B_{p/2} \\ 2C_{p/2}D_{p/2} & A_{p/2}D_{p/2} + B_{p/2}C_{p/2} \end{bmatrix} \tag{5}$$

and the corresponding dispersion relation can be written as

$$\cosh(\gamma p) = A_{p/2}D_{p/2} + B_{p/2}C_{p/2} = 2A_{p/2}D_{p/2} - 1 \tag{6}$$

(taking into account the general condition $AD - BC = 1$). If we now consider the identity

$$\cosh(\gamma p) = 2\cosh^2(\gamma p/2) - 1 \tag{7}$$

it can be concluded that

$$\cosh(\gamma p/2) = \sqrt{A_{p/2}D_{p/2}}. \tag{8}$$

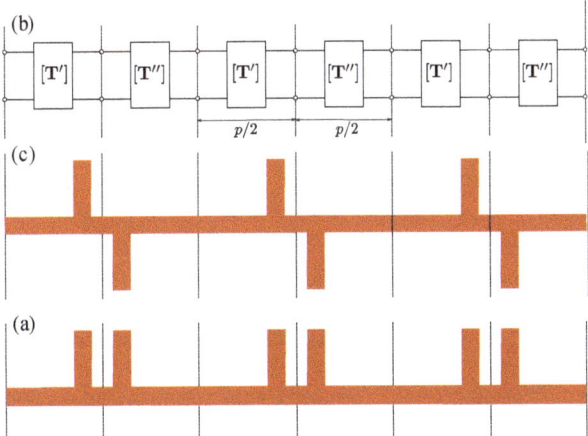

Figure 2. Periodic structure (period $\equiv p$) with a symmetry plane at the middle of the unit cell. (**a**) Cascade of 2-port equivalent networks that explicitly takes into account the presence of the symmetry plane in the unit cell; (**b**) Structure with a point of inversion symmetry; (**c**) Structure with a plane of reflection symmetry.

This dispersion equation is exactly the same as the one given in ([41] Equation (6)) (note that the minimum period was denoted as $2p$ in [41] and p in this work). In principle we can compute the product $A_{p/2}D_{p/2}$ from the scattering parameters provided by the full-wave simulation of half the unit cell when considered isolated. However, if we proceed this way, all the higher-order interactions between the two halves of the unit cell would be ignored. (The two-port transmission matrices $[\mathbf{T}']$ and $[\mathbf{T}'']$ should rather be considered again as "effective" transmission matrices, in the understanding that the fundamental mode might not be the only one that contributes to the interaction between the two halves of the unit cell.)

Alternatively we could have proceeded in the manner reported in [41] by introducing a short/open circuit in the structure in order to compute $A_{p/2}D_{p/2}$ in terms of the auxiliary impedances Z_{el} and Z_{mag} defined in [41]. These are the input impedances of the one-port network obtained by substituting the symmetry plane located at the center of the unit cell (see Figure 2) with a short circuit (Z_{el}) or an open circuit (Z_{mag}). This procedure is found equivalent to starting with the whole unit cell and then applying the even/odd excitation technique [3,4]. As this method implies the setting of electric/magnetic walls at the middle of the structure, both procedures are fully equivalent provided the actual existence of a reflection symmetry plane in the structure that allows for the application of the even/odd excitation procedure. As an example, this symmetry condition is satisfied for the geometry given in Figure 2c but not for the one in Figure 2b. This equivalence, or lack of equivalence, could be irrelevant if it were not for the fact that the presence of electric/magnetic walls in the full-wave simulation is what actually ensures that the higher-order interaction between the two symmetric halves of the unit cell is appropriately taken into account. It is also apparent that the use of these magnetic/electric walls does not imply that the higher-order interaction between *adjacent* whole unit cells is taken into account, since only one unit cell is indeed analyzed with the full-wave simulator. Moreover, the placing of a magnetic wall in the input port of the half unit cell imposed in [41] does not affect this discussion; actually this magnetic wall is not necessary when the input lumped port is taken at the middle of the unit cell [41] (this fact has repeatedly been checked by many numerical simulations carried out by the authors of the present work). In brief, the technique reported in [41] can, in principle, be applied to geometries of the type shown in Figure 2b,c, but the application of that technique to the geometry in Figure 2b would only account for the electromagnetic interactions between the two halves of the unit cell as long as this interaction is carried out exclusively by the fundamental mode (in other words, the two stubs in each unit cell should be sufficiently far apart so as to ensure the absence of interactions through high-order modes). It leads us to the somehow trivial finding that the higher-order interaction between adjacent cells can only be taken into account by simulating a cascade of multiple unit cells (as done in Method A) or, at most, by taking advantage of reflection symmetry planes (not inversion points, as it is the case shown in Figure 2b) to simulate half the cascade of unit cells terminated with electric/magnetic walls. This result will be numerically studied and validated in next section.

A general conclusion drawn above was that only method A can properly account for high-order interactions between different unit cells, and thus method B should be restricted to those situations where this interaction can be neglected. However, a possible and convenient way to account for the high-order interactions between adjacent cells, which takes advantage of the combination of methods A and B, is to apply method B to an extended cell of period $P = 2p$. In this case, the ABCD parameters in (8) would correspond just to the unit cell of period p, which because of the presence of a reflection-symmetry wall in the middle of the extended cell ($A_p = D_p$) would lead to a dispersion equation completely equivalent to (3) if N had been set to 2 in method A. Certainly this combined procedure can be applied to cells of more extended periodicity with the purpose of accounting for inter-cell interactions with simulations that only involve half the number of cells.

Interestingly, the above discussion about symmetry in periodic structures turns out to be very relevant when dealing with the circuit modeling of structures with glide/twisted symmetry [44,45]. This topic has recently surged due to some interesting application papers [46–49], where it is clearly shown that the behavior of periodic structures with glide symmetry is not equivalent to their counterpart without this feature, thus giving an apparent clue on the different role played by the higher-order mode coupling when different types of symmetry are involved. Furthermore, this difference in coupling provided by the glide symmetry has been found to give advantageous features that can enhance the performance of many practical devices [46,48,49]. According to our discussion above, the periodic structure with glide symmetry (whose unit cell incorporates a central inversion point) cannot rigorously be modeled by the analysis of just one of the two subcells of the unit cell (unlike periodic structures whose unit cell does have a reflection symmetry plane). Actually, the authors of [47] claim that their circuit model is valid provided that the upper and lower stubs in ([47], Figure 2c,d) do not overlap. Our premise here is that their proposed simplified circuit modeling of the glide-symmetric structure is valid as long as the upper/lower position of the stubs is irrelevant; namely, when the higher-order coupling between the stubs is not very important. When both stubs overlap this possible difference in the higher-order coupling between upper/lower position is crucial, being less and less relevant as the distance between the stubs increases. In practice, as already reported in [50], there may be many practical situations where just the inclusion of the first high-order mode suffices to obtain accurate results.

Also, the above discussion can be related to a very recent contribution in the circuit modeling of non-symmetrical reciprocal network [51]. Although that paper deals with non-periodic structures, its underlying rationale can easily be extended to the periodic case, in which the general conclusions reached in [51] are found congruent with the discussions reported here.

3. Results

In this section the main issues discussed in previous sections will be numerically validated. First, the general advantage of using Method B will be pointed out when possible. Certainly Method A will provide, in principle, more accurate results since it deals with a more realistic electromagnetic scenario in which many of the couplings between different unit cells are taken into account. However, the unavoidable computational load implicit in the treatment of electrically large and complex structures may lead to very long computational times and non-negligible levels of numerical noise. This last effect can become very relevant when dealing, for instance, with leaky-wave 1-D periodic configurations, where the eventual high radiation leakage in the structure can make the power in the output port several orders of magnitude smaller than in the input port. This numerical noise is also very relevant when computing the attenuation constant of below-cutoff and/or complex modes in closed waveguiding system. A few selected examples will be discussed in the next subsections to clarify these points and provide some insights on the virtues and limitations of the proposed methods.

3.1. Periodic Printed Microstrip Lines

Our first case study in Figure 3 shows the comparison between the results of methods A and B for two printed microstrip lines periodically loaded with inductive/capacitive discontinuities. These lines were previously studied in [38,52] and, in this example, HFSS commercial software [29] has been used for the required full-wave simulations. Both structures exhibit a band gap starting when its period p equals half the line wavelength ($\lambda_g/2$) (as shown in [35,38]). Although not explicitly shown in the figures, our results agree well with those reported in [38,52].

In Figure 3a the results for the dispersion diagram (both the phase, β, and the attenuation, α, constants) corresponding to method A have been computed using one cell ($N = 1$) and five cells ($N = 5$). In this structure, the differences between the $N = 1$ and $N = 5$ cases are very small, clearly meaning that the inter-cell coupling is well accounted for by just the fundamental mode. The data corresponding to method B have been obtained by using the even-odd excitation procedure to study the unit cell (namely, just one of the two symmetric halves with electric/magnetic walls are simulated). Certainly, the results using this procedure are found identical to the ones obtained by the procedure proposed in ([41], Equation (12)) as well as to the "Method A $N = 1$" curve. The excellent agreement between the "Method B" curve and the results with $N = 1$ can be considered as a first validation of the congruence of methods A and B in this circumstance.

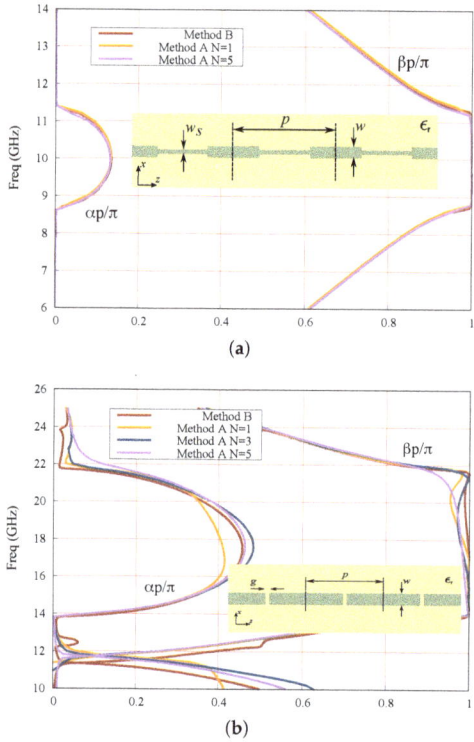

Figure 3. Dispersion diagram of the periodic printed microstrip lines shown as insets. The substrate of the lines has thickness h and relative permittivity ϵ_r. (**a**) $p = 5.6$ mm, $w = 0.6$ mm, $w_s = 0.2$ mm, $\epsilon_r = 10.2$ and $h = 0.635$ mm; (**b**) $p = 4$ mm, $w = 0.6$ mm , $g = 0.2$ mm, $\epsilon_r = 10.2$ and $h = 0.767$ mm.

In the structure analyzed in Figure 3b, previously studied in [52], the results provided by method A for one, three, and five cells (denoted as $N = 1$, $N = 3$, and $N = 5$ respectively) are compared with those obtained by method B, taking now an extended period of $2p$ (namely, the simulated subcell with electric/magnetic walls has a length of p, and thus the obtained results are found to be identical to the case "Method A $N = 2$", although this fact is not explicitly shown in the figure since the two corresponding curves would overlap). It is interesting to observe the appearance of slight discrepancies between the results of method A when different number of cells are considered and, furthermore, that a clear convergence pattern is not observed in the analyzed frequency range. These facts are partly attributable to the high values of the attenuation constant in the stopband (there appear values of $\alpha/k_0 > 1$), which causes the results with a few cells to be affected by numerical noise due to the very low level of power that reaches the output port (power along z is given by $P(z) = P(0)e^{-2\alpha z}$; that is, an attenuation of $\mathcal{A}(\text{dB}) = 8.69\alpha z$, which means $\mathcal{A} = 13.11 N$ dB at 18 GHz, with N being the number of cells in the structure). In this situation, the convenient convergence test to ensure the reliability of the results provided by method A cannot be carried out. Actually, although not explicitly shown in Figure 3b, the results with a higher number of cells are found to increasingly deteriorate.

3.2. Corrugated Parallel-Plate Waveguide

The following example to be examined is a parallel-plate waveguide (PPW) system with periodic metallic corrugations, which can be symmetrically and non-symmetrically distributed, as illustrated by the insets in Figure 4a,b respectively (here the term "symmetrically distributed" is employed to refer only to "reflection symmetry"). First, Figure 4 shows the case where the period of the structure is sufficiently long as to make the inter-cell coupling due to high-order modes almost negligible up to the cutoff frequency of the first high-order mode of the housing PPW. In this frequency range the use of two-port ABCD matrices is well justified and thus the present structure will be taken as a good benchmark to study the intra-cell high-order couplings and its relation to the symmetry properties of the unit cell. In this long period case, no method-A results with $N > 1$ are shown since they are found to almost coincide with the $N = 1$ case (although some small numerical noise appears because of the inherent more difficult simulation of electrically large structures). Our results will be compared with the data provided by the tool "Eigenmode-Solver" in CST [28]. The results provided by this tool for the frequency behavior of the phase constant are considered very reliable; however, as this tool does not straightforwardly generate results for the attenuation constant of reactive/complex/leaky modes, the comparison for α is not carried out.

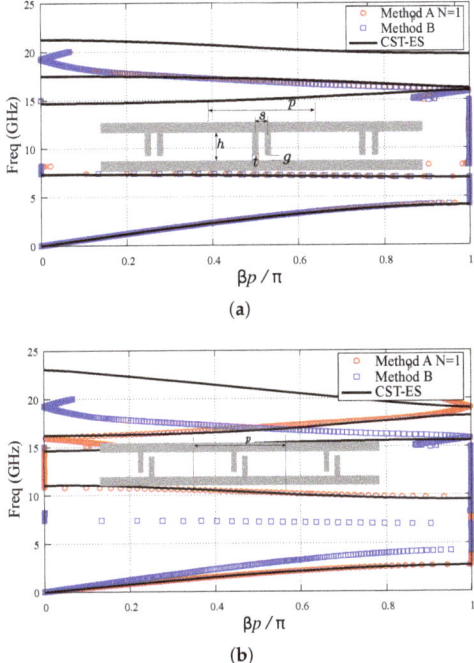

Figure 4. Brillouin diagram of a parallel-plate waveguide (PPW) with periodic metallic corrugations (**a**) symmetrically and (**b**) non-symmetrically distributed. The inset structures represent longitudinal sections of the PPW's with the grey color standing for the metallic parts of the structures. Long period: $p = 12$ mm. Other parameters: $h = 10$ mm, $s = 2$ mm, $t = 1$ mm and $g = 1$ mm.

The reflection-symmetry case studied in Figure 4a clearly shows a very good agreement between the results of method A with just one period ($N = 1$) and those given by the even-odd excitation procedure in method B in the whole considered frequency range. These results agree well with the data provided by the CST Eigenmode-Solver (CST-ES) up to 15 GHz; namely, the cutoff frequency of the first high-order mode in the housing *uniform* (without corrugations) PPW: $f = c/(2h)$, with c

being the vacuum speed of light and h the height of the uniform PPW. As expected, discrepancies start to appear beyond this frequency, when the presence of this second propagative mode in the uniform sections of the PPW makes that the cascade of two-port ABCD matrices cannot properly account for the unavoidable multi-mode inter-cell coupling that will appear. It should be highlighted that, for frequencies where the regime is no longer monomode in the non-corrugated PPW, the cascade of single two-port transfer matrices is not presumed to work satisfactorily because of the *propagative* multi-mode nature of the guiding field rather than for possible neglected effects of high-order *evanescent* modes. Thus, beyond this monomode band, neither method A nor method B are expected to provide accurate results.

Let us now consider the inversion-symmetry structure shown in the inset of Figure 4b. The structure exhibits an inversion point at the center of the unit cell, circumstance that will have interesting consequences for a particular choice of the period, as it will be discussed later. The curves in Figure 4b show again a good agreement within the monomode band between the results of method A and those provided by the CST-ES. However, the data given by method B, which here completely ignores the inversion-symmetry nature of the unit cell, are in full disagreement with the above two set of results. It clearly proves that method B, as expected, drastically fails when high-order coupling between the two subcells forming the inversion-symmetry unit cell is significant and not properly incorporated to the model. This fact seems to be in contrast with the thesis apparently sustained in [41]. Of course, the method would properly work if the subcells only interact through the fundamental propagating mode, as it would be the case for relatively large separation between the stubs. In this latter case the structures with mirror symmetry and with inversion symmetry would have the same response.

Next, the short-period case is studied in Figure 5, where it is again shown the cases corresponding to the presence or absence of a reflection symmetry plane at the middle of the unit cell. In both cases it is now included the data for "Method A $N = 5$" since high-order inter-cell coupling is expected when the pair of corrugations of each unit cell is electrically close to adjacent ones. Figure 5a shows the symmetrically-distributed case and, again, a perfect agreement is found between "Method A $N = 1$" and "Method B". These methods also show a good agreement with the "CST-ES" data in the first passband up to 5 GHz where $p/\lambda_0 \lesssim 1/10$. For higher frequencies, the inter-cell couplings cannot be well accounted for by the cascade of two-port ABCD matrices and, therefore, the methods are not expected to give accurate quantitative results (although they still provide a good qualitative picture of the band diagram of the structure). The data corresponding to "Method A $N = 5$" shows a significantly better agreement with the "CST-ES" curve in the whole first passband, although important discrepancies appear in the stopband due to the expected numerical noise caused by the strong attenuation in this band. There is neither a good agreement in the second passband because of the multi-mode nature of the band at higher frequencies. Due to the unreliability of these "Method A $N = 5$" data outside the first passband, they are not shown in the figure. Regarding the inversion-symmetry case in Figure 5b, the first relevant feature is the complete lack of agreement between the results of "Method A $N = 1$" and "Method B" even at very low frequencies, which clearly highlights the need of appropriately considering the multi-mode interactions that appear here between the two halves of the unit cell. Since the first passband in this case appears below 5 GHz, a very good agreement is now found between the "Method A $N = 1$" and "CST-ES". For higher frequencies there is not such a good quantitative agreement, although the qualitative behavior is approximately captured until the onset of the multi-mode propagation regime. In this inversion-symmetry case, the "Method A $N = 5$" curve does show a good quantitative agreement within the entire single-band regime, agreement that extends up to the onset of the multi-mode regime. An interesting feature of this structure is the widening of the second bandpass of negative-group velocity nature [53], which now extends from 8 to 13 GHz approximately.

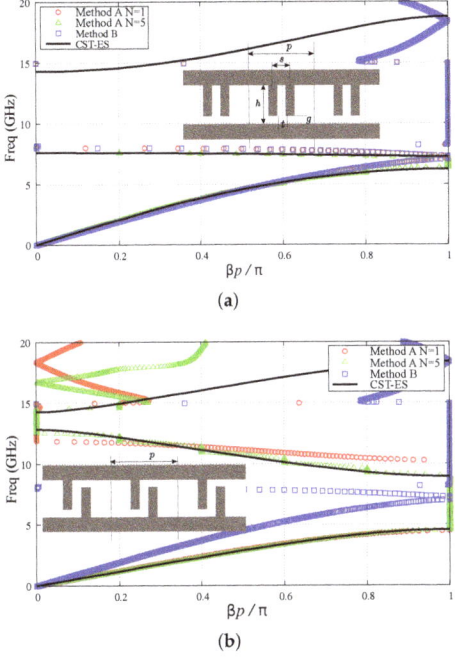

Figure 5. Brillouin diagram of a parallel-plate waveguide with periodic metallic corrugations (**a**) symmetrically and (**b**) non-symmetrically distributed. Short period: $p = 6$ mm. Other parameters: $h = 10$ mm, $s = 2$ mm, $t = 1$ mm and $g = 1$ mm.

An interesting further step in the study of corrugated PPWs is the case presented in Figure 6, where the period of the structure is taken as twice the distance between the corrugations [47]. Actually, for the symmetrically-distributed corrugation case shown in Figure 6a, the true period of the structure is now $p = s$, although the dispersion relation will be plotted for an "extended" period $\hat{p} = 2s$ for the sake of comparison with the inversion-symmetry case in Figure 6b [the true period of which is $p = 2s$]. The CST-EC curve in Figure 6a shows a first passband that now extends up to about 8 GHz; interestingly, the onset of the second bandstop in Figure 5a. Actually, it is found that the first stopband in this structure disappears as the distance between corrugations approaches the true period of the structure ($p \to s$). In this figure it is also observed that the "Method A N = 2" gives sufficiently accurate quantitative results in this first passband but drastically fails for higher frequencies (in these calculations, a structure with two unit cells, $\hat{p} = 2s$, has been taken). The curve corresponding to "Method A N = 10" only improves the agreement with CST-ES in the first passband but also fails for higher frequencies (the data are not shown). In this case, the expected relevant inter-cell propagative and evanescent high-order coupling would make it necessary either the use of a multi-port approach or the solution of the corresponding non-linear eigenvalue problem to compute the complex propagation constant of the structure. The case considered in Figure 6b is an example of a structure having glide symmetry [46–49]. The curves of Figure 6b show, similarly to the previous inversion-symmetry unit-cell cases, a great disagreement between the results of Method A (N = 1) and Method B. Rather interesting is the comparison of the "Method A N = 5" and "CST-ES" curves. Now there appears a good agreement between both curves even for frequencies beyond 15 GHz. This good agreement outside the monoband regime is attributed to the fact that the first high-order mode of the housing PPW is not expected to be highly excited in the glide-symmetric structure. The existence of electrically-close periodic inversion points in this structure would presumably reduce the excitation level of modes with an even profile along the vertical direction. Apart from this reasonable good agreement in the quantitative results shown by both curves, they show that the passband of this structure has grown considerably, now extending up to 13.8 GHz. This surprising fact, already reported in [46–49], is one of the most relevant features of glide-symmetric structures and is likely to find more and more applications in the future (a similar widening of a stopband can also be found in other type of glide-symmetric structures).

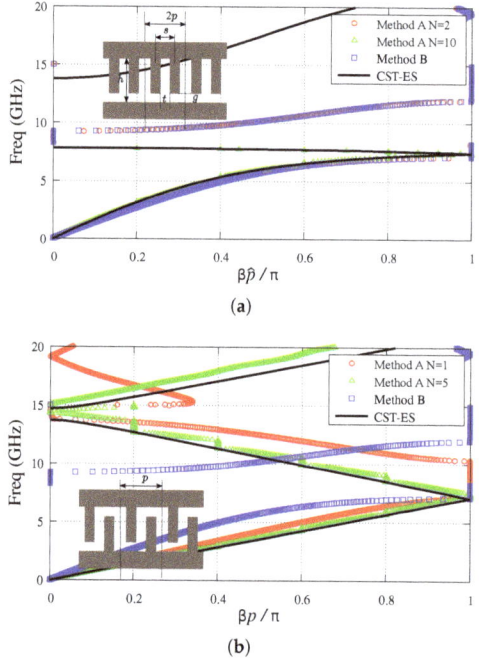

Figure 6. Brillouin diagram of a parallel-plate waveguide with periodic metallic corrugations (**a**) symmetrically [$\hat{p} = 2p = 4$ mm] and (**b**) glide-symmetrically distributed [$p = 4$ mm]. Other parameters: $h = 10$ mm, $s = 2$ mm, $t = 1$ mm and $g = 1$ mm.

4. Conclusions

In this work we have presented a thorough discussion on the pros and cons of computing the dispersion relation of 1-D periodic guiding/radiating structures by means of a combined method that makes use of full-wave simulations data obtained from commercial software tools along with a simplifying equivalent-network model. This method can take advantage of both the high flexibility of simulators to deal with general structures and the further analytical treatment of the data provided by the employed simplified electromagnetic model of the problem as a cascade of two-port equivalent networks.

Some general conclusions that have been discussed in this work are summarized below:

- The most general and reliable method to compute the dispersion diagram of periodic structures is the solution of Maxwell's equations in the unit cell of the structure subject to periodic boundary conditions. The major disadvantage of this eigenvalue approach is that it involves the searching for complex zeros, a task that is not easily systematized in the form of general-purpose computer codes.
- The so-called Methods A and B in Section 2 are alternative combined approaches that avoid the searching of complex zeros. If the unit cell of the periodic structure has a reflection-symmetry

- plane, then a further combination of these two methods is proposed as a very convenient and efficient tool.
- Due to the simplification implicit in the modeling of the periodic structure as a cascade of two-port equivalent networks, the possible excitation of high-order propagative modes in the uniform sections of the housing structure will restrict the application of this technique to the so-called monoband regime.
- The combined technique would need to deal with structures with many unit cells in order to take into account the possible inter-cell coupling due to high-order evanescent modes. When inter-cell coupling due to either propagative or evanescent high-order modes can be neglected, this method is always expected to work properly.
- In contrast with the above fact, some authors have claimed that the convenient treatment of just one unit cell would suffice if a proper even-odd periodic excitation technique is applied. In this work it is discussed that this methodology can be used reliably only when there is a reflection-symmetry plane in the unit cell or the interaction between the two halves of the unit cell is carried out by only the fundamental mode.
- Finally, it has also been discussed that glide-symmetric structures, whose interesting properties have recently been the object of intensive study, will require the modeling of the unit cell as a multi-port equivalent network or the solution of the corresponding rigorous eigenvalue problem.

Author Contributions: The contributions of all of the authors were the same. All of them have worked together to develop the present manuscript.

Funding: This research was funded by [the Spanish Ministerio de Ciencia, Innovación y Universidades with European Union FEDER funds] grant number [TEC2017-84724-P].

Acknowledgments: The authors would like to express their gratitude to Alejandro Martínez-Ros for his contribution in the first steps of the present work.

Conflicts of Interest: The authors declare no conflict of interest.

Abbreviations

The following abbreviations are used in this manuscript:

CST	Computer Simulation Technology
FSS	Frequency Selective Surface(s)
HFSS	High Frequency Structure Simulator
PPW	Parallel Plate Waveguide

References

1. Montgomery, C.G.; Dicke, R.H.; Purcell, E.M. *Principles of Microwave Circuits*; MIT Radiation Laboratory Series; McGraw-Hill: New York, NY, USA, 1948; Volume 8.
2. Marcuvitz, N. *Waveguide Handbook*; MIT Radiation Laboratory Series; McGraw-Hill: New York, NY, USA, 1951; Volume 10.
3. Collin, R. *Field Theory of Guided Waves*; McGraw Hill: New York, NY, USA, 1960.
4. Pozar, D.M. *Microwave Engineering*, 3rd ed.; Wiley: Hoboken, NJ, USA, 2005.
5. Cameron, R.J.; Kudsia, C.M.; Mansour, R.R. *Microwave Filters for Communication Systems*; Wiley: Hoboken, NJ, USA, 2007.
6. Eleftheriades, G.V.; Balmain, K.G. *Negative-Refraction Metamaterials: Fundamental Properties and Applications*; Wiley: Hoboken, NJ, USA, 2005.
7. Martín, F. *Artificial Transmission Lines for RF and Microwave Applications*; Wiley: Hoboken, NJ, USA, 2015.
8. Jackson, D.R.; Oliner, A.A. Leaky-wave antennas. In *Modern Antenna Handbook*; John Wiley and Sons: Hoboken, NJ, USA, 2007; Chapter 7, pp. 325–367.
9. Munk, B. *Frequency Selective Surfaces: Theory and Design*; John Wiley and Sons: Hoboken, NJ, USA, 2000.
10. Huang, J.; Encinar, J.A. *Reflectarray antennas*; Wiley, Inter Science: Hoboken, NJ, USA, 2007.

11. Gagnon, N.; Petosa, A.; McNamara, D.A. Research and development on phase-shifting surfaces (PSSs). *IEEE Antennas Propag. Mag.* **2013**, *55*, 29–48. [CrossRef]
12. Martini, E.; Maci, S. Metasurface Transformation Theory. In *Transformation Electromagnetics and Metamaterials*; Springer: London, UK, 2014; pp. 83–116.
13. Kurokawa, K. *An Introduction to the Theory of Microwave Circuits*; Academic Press: San Francisco, CA, USA, 1969.
14. Varela, J.E.; Esteban, J. Characterization of waveguides with a combination of conductor and periodic boundary contours: Application to the analysis of bi-periodic structures. *IEEE Trans. Microw. Theory Tech.* **2012**, *60*, 419–430. [CrossRef]
15. Rodríguez-Berral, R.; Mesa, F.; Medina, F. Analytical multimodal network approach for 2-D arrays of planar patches/apertures embedded in a layered medium. *IEEE Trans. Antennas Propag.* **2015**, *63*, 1969–1984. [CrossRef]
16. Mesa, F.; Rodríguez-Berral, R.; Medina, F. Unlocking complexity with ECA. *IEEE Microw. Mag.* **2018**, *19*, 44–65. [CrossRef]
17. Hong, J.S. *Microstrip Filters for RF/Microwave Applications*, 2nd ed.; Wiley: Hoboken, NJ, USA, 2011.
18. Feng, X.; Ke, W. Guided-wave and leakage characteristics of substrate integrated waveguide. *IEEE Trans. Microw. Theory Tech.* **2005**, *33*, 66–73. [CrossRef]
19. Bozzi, M.; Pasian, M.; Perregrini, L.; Wu, K. On the losses in substrate integrated waveguides and cavities. *Int. J. Microw. Wirel. Technol.* **2009**, *1*, 395–401. [CrossRef]
20. Rubio, J.; Gómez-García, A.; Gómez-Alcalá, R.; Campos-Roca, Y.; Zapata, J. Overall formulation for multilayer SIW circuits based on addition theorems and the generalized scattering matrix. *IEEE Microw. Wirel. Compon. Lett.* **2018**, *28*, 485–487. [CrossRef]
21. Collin, R.E.; Zucker, F.J.; (Eds) *Antenna Theory*; McGraw-Hill: New York, NY, USA, 1969.
22. Peterson, A.F.; Ray, S.L.; Mittra, R. *Computational Methods for Electromagnetics*; IEEE Press: New York, NY, USA, 1998.
23. Dudley, D.G. *Mathematical Foundations for Electromagnetic Theory*; IEEE Press: New York, NY, USA, 1994.
24. Felsen, L.B.; Marcuvitz, N. *Radiation and Scattering of Waves*; Prentice-Hall: Upper Saddle River, NJ, USA, 1973.
25. Rodríguez-Berral, R.; Mesa, F.; Medina, F. Systematic and efficient root finder for computing the modal spectrum of planar layered waveguides. *Int. J. RF Microw. Comput. Eng.* **2004**, *14*, 73–83. [CrossRef]
26. Kowalczyk, P. Complex Root Finding Algorithm Based on Delaunay Triangulation. *ACM Trans. Math. Softw.* **2015**, *41*, 19. [CrossRef]
27. Zouros, G.P. CCOMP: An efficient algorithm for complex roots computation of determinantal equations. *Comput. Phys. Commun.* **2018**, *222*, 339–350. [CrossRef]
28. CST Microwave Studio. 2017. Available online: https://www.cst.com/products/cstmws (accessed on July 2018).
29. ANSYS High Frequency Structure Simulator (HFSS). Available online: https://www.ansys.com/products/electronics/ansys-hfss (accessed on July 2018).
30. Sampath, M.K. On addressing the practical issues in the extraction of RLGC parameters for lossy multiconductor transmission lines using S-parameter models. In Proceedings of the Electrical Performance of Electronic Packaging (EPEP), San Jose, CA, USA, 27–29 October 2008; pp. 259–262.
31. Apaydin, N.; Zhang, L.; Sertel, K.; Volakis, J.L. Experimental validation of frozen modes guided on printed coupled transmission lines. *IEEE Trans. Microw. Theory Tech.* **2012**, *60*, 1513–1519. [CrossRef]
32. Liu, Z.; Zhu, L.; Wu, Q.; Xiao, G. A short-open calibration (SOC) technique to calculate the propagation characteristics of substrate integrated waveguide. In Proceedings of the 2015 IEEE MTT-S International Microwave Workshop Series on Advanced Materials and Processes for RF and THz Applications, IEEE MTT-S IMWS-AMP, Suzhou, China, 1–3 July 2015.
33. Conciauro, G.; Guglielmi, M.; Sorrentino, R. *Advanced Modal Analysis*; Wiley: Hoboken, NJ, USA, 1999.
34. Esteban, J.; Rebollar, J.M. Characterization of corrugated waveguides by modal analysis. *IEEE Trans. Microw. Theory Tech.* **1991**, *39*, 937–943. [CrossRef]
35. Baccarelli, P.; Di Nallo, C.; Paulotto, S.; Jackson, D.R. A full-wave numerical approach for modal analysis of 1D periodic microstrip structures. *IEEE Trans. Microw. Theory Tech.* **2006**, *54*, 1350–1362. [CrossRef]

36. Paulotto, S.; Baccarelli, P.; Frezza, F.; Jackson, D.R. Full-Wave modal dispersion analysis and broadside optimization for a class of microstrip CRLH leaky-wave antennas. *IEEE Trans. Microw. Theory Tech.* **2008**, *56*, 2826–2837. [CrossRef]
37. Mao, S.G.; Chen, M.Y. Propagation characteristics of finite-width conductor-backed coplanar waveguides with periodic electromagnetic bandgap cells. *IEEE Trans. Microw. Theory Tech.* **2002**, *50*, 2624–2628.
38. Zhu, L. Guided-wave characteristics of periodic microstrip lines with inductive loading: Slow-wave and bandstop behaviors. *Microw. Opt. Technol. Lett.* **2004**, *41*, 77–79. [CrossRef]
39. Valerio, G.; Paulotto, S.; Baccarelli, P.; Burghignoli, P.; Galli, A. Accurate Bloch analysis of 1-D periodic lines through the simulation of truncated structures. *IEEE Trans. Antennas Propag.* **2011**, *59*, 2188–2195. [CrossRef]
40. Martínez-Ros, A.J.; Mesa, F. A study on the dispersion relation of periodic structures using commercial simulators. In Proceedings of the 2017 Computing and Electromagnetics International Workshop (CEM), Barcelona, Spain, 21–24 June 2017; pp. 15–16.
41. Eberspacher, M.A.; Eibert, T.F. Dispersion analysis of complex periodic structures by full-wave solution of even-odd-mode excitation problems for single unit cells. *IEEE Trans. Antennas Propag.* **2013**, *61*, 6075–6083. [CrossRef]
42. Monni, S.; Gerini, G.; Neto, A.; Tijhuis, A.G. Multi-mode equivalent networks for the design and analysis of frequency selective surfaces. *IEEE Trans. Antennas Propag.* **2007**, *55*, 2824–2835. [CrossRef]
43. Kaipa, C.S.R.; Yakovlev, A.B.; Medina, F.; Mesa, F. Transmission through stacked 2-D periodic distributions of square conducting patches. *J. Appl. Phys.* **2012**, *112*, 033101. [CrossRef]
44. Hessel, A.; Oliner, A.A.; Chen, M.; Li, R. Propagation in periodically loaded waveguides with higher symmetries. *Proc. IEEE* **1973**, *61*, 183–195. [CrossRef]
45. Amari, S.; Vahldieck, R.; Bornemann, J. Accurate analysis of periodic structures with an additional symmetry in the unit cell from classical matrix eigenvalues. *IEEE Trans. Microw. Theory Tech.* **1998**, *46*, 1513–1515. [CrossRef]
46. Quevedo-Teruel, O.; Ebrahimpouri, M.; Kehn, M.N.M. Ultra-wideband metasurface lenses based on off-shifted opposite layers. *IEEE Antennas Wirel. Propag. Lett.* **2016**, *15*, 484–487. [CrossRef]
47. Valerio, G.; Sipus, Z.; Grbic, A.; Quevedo-Teruel, O. Accurate equivalent-circuit descriptions of thin glide-symmetric corrugated metasurfaces. *IEEE Trans. Antennas Propag.* **2017**, *65*, 2695–2700. [CrossRef]
48. Dahlberg, O.; Mitchell-Thomas, R.; Quevedo-Teruel, O. Reducing the dispersion of periodic structures with twist and polar glide symmetries. *Sci. Rep.* **2017**, *7*, 10136. [CrossRef] [PubMed]
49. Ebrahimpouri, M.; Rajo-Iglesias, E.; Sipus, Z.; Quevedo-Teruel, O. Cost-Effective gap waveguide technology based on glide-symmetric holey EBG structures. *IEEE Trans. Microw. Theory Tech.* **2018**, *6*, 927–934. [CrossRef]
50. Naqui, J.; Duran-Sindreu, M.; Fernandez-Prieto, A.; Mesa, F.; Medina, F.; Martin, F. Multimode propagation and complex waves in CSRR-based transmission-line metamaterials. *IEEE Antennas Wirel. Propag. Lett.* **2012**, *11*, 1024–1027. [CrossRef]
51. Abdo-Sanchez, E.; Camacho-Peñalosa, C.; Martin-Guerrero, T.; Esteban, J. Equivalent circuits for non-symmetric reciprocal two-ports based on eigen-state formulation. *IEEE Trans. Microw. Theory Tech.* **2017**, *65*, 4812–4822. [CrossRef]
52. Rodriguez-Berral, R.; Mesa, F.; Baccarelli, P.; Burghignoli, P. Excitation of a periodic microstrip line by an aperiodic delta-gap source. *IEEE Antennas Wirel. Propag. Lett.* **2009**, *8*, 641–644. [CrossRef]
53. Hwang, R. Negative group velocity and anomalous transmission in a one-dimensionally periodic waveguide. *IEEE Trans. Antennas Propag.* **2006**, *54*, 755–760. [CrossRef]

© 2018 by the authors. Licensee MDPI, Basel, Switzerland. This article is an open access article distributed under the terms and conditions of the Creative Commons Attribution (CC BY) license (http://creativecommons.org/licenses/by/4.0/).

Article

On the Electric-Magnetic Duality Symmetry: Quantum Anomaly, Optical Helicity, and Particle Creation

Iván Agulló [1], Adrián del Río [2] and José Navarro-Salas [3,*]

1. Department of Physics and Astronomy, Louisiana State University, Baton Rouge, LA 70803-4001, USA; agullo@lsu.edu
2. Centro de Astrofísica e Gravitacâo—CENTRA, Departamento de Física, Instituto Superior Técnico—IST, Universidade de Lisboa, 1049 Lisboa, Portugal; adriandelrio@tecnico.ulisboa.pt
3. Departamento de Física Teórica and IFIC, Centro Mixto Universidad de Valencia-CSIC, Facultad de Física, Universidad de Valencia, 46100 Burjassot, Valencia, Spain
* Correspondence: jnavarro@ific.uv.es

Received: 14 November 2018; Accepted: 13 December 2018; Published: 17 December 2018

Abstract: It is well known that not every symmetry of a classical field theory is also a symmetry of its quantum version. When this occurs, we speak of quantum anomalies. The existence of anomalies imply that some classical Noether charges are no longer conserved in the quantum theory. In this paper, we discuss a new example for quantum electromagnetic fields propagating in the presence of gravity. We argue that the symmetry under electric-magnetic duality rotations of the source-free Maxwell action is anomalous in curved spacetimes. The classical Noether charge associated with these transformations accounts for the net circular polarization or the optical helicity of the electromagnetic field. Therefore, our results describe the way the spacetime curvature changes the helicity of photons and opens the possibility of extracting information from strong gravitational fields through the observation of the polarization of photons. We also argue that the physical consequences of this anomaly can be understood in terms of the asymmetric quantum creation of photons by the gravitational field.

Keywords: electric-magnetic duality symmetry; quantum anomalies; optical helicity; electromagnetic polarization; particle creation

---

1. Introduction

Symmetries are at the core of well-established physical theories, and they keep playing a central role in the mainstream of current research. Fundamental Lagrangians in physics are founded on symmetry principles. Moreover, symmetries are linked, via Noether's theorem, to conservations laws. Well-known examples are the energy and momentum conservation and its relation with the invariance under space–time translations, as well as the conservation of the net fermion number (the difference in the number of fermions and anti-fermions that is proportional to the net electric charge) in Dirac's relativistic theory, which result from the global phase invariance of the action.

When the symmetries of free theories are also preserved by interactions, the conservation laws are maintained, and they can be used to understand patterns in diverse physical phenomena. In quantum electrodynamics, for instance, the phase invariance is preserved by the coupling of the Dirac and the electromagnetic field, and this ensures the conservation of the net fermion number in all physical processes [1]. Another illustrative example is the gravitationally induced creation of particles, either bosons or fermions, in an expanding homogenous universe [2–5]. This particle creation occurs in pairs, and the symmetry of the background under space-like translations ensures that, if one particle is created with wavenumber $\vec{k}$, its partner has wavenumber $-\vec{k}$. As a consequence, there is no creation of net momentum, as expected on symmetry grounds. In a similar way, phase invariance implies that the gravitational field cannot create a net fermion number in an expanding universe.

However, in special cases, the implications of classical symmetries do not extend to quantum theory, and the classical charge conservation breaks down. This was first noticed by studying massless fermions coupled to an electromagnetic field [6,7]. A massless fermion is called a (Weyl) left-handed fermion if it has helicity $h = -1/2$, and a right-handed fermion if $h = +1/2$ (A left-handed (right-handed) anti-fermion has helicity $h = +1/2$ $(-1/2)$). Recall that the equations of motion for the two sectors decouple in the massless limit, and this allows one to write a theory for massless fermions that involves only one of the two helicities, something that is not possible for non-zero mass. The action of this theory also enjoys phase invariance, so the number of left-handed and right-handed fermions is *separately* conserved. This is to say, in the classical theory, there are two independent Noether currents, j_L^μ and j_R^μ, associated with left- and right-handed massless fermions, respectively, that satisfy continuity equations $\partial_\mu j_L^\mu = 0$ and $\partial_\mu j_R^\mu = 0$. Rather than using j_L^μ and j_R^μ, it is more common to re-write these conservation laws in terms of the so-called vector and axial currents, defined by their sum and difference, respectively, $j^\mu = j_R^\mu + j_L^\mu = \bar{\psi}\gamma^\mu\psi$ and $j_5^\mu = j_R^\mu - j_L^\mu = \bar{\psi}\gamma^\mu\gamma^5\psi$, where ψ is the four-component Dirac spinor, that encapsulates both left- and right-handed (Weyl) fermions, and γ^μ, γ_5 are the Dirac matrices.

What is the situation in quantum theory? It turns out that the conservation law for j^μ holds also quantum mechanically, so the quantum number $N_R + N_L$ (associated with the net fermion number, i.e., the electric charge) is preserved in any physical process. For instance, in the presence of a time-dependent electromagnetic background, charged fermions and antifermions are spontaneously created (this is the electromagnetic analog of the gravitationally induced particle creation mentioned above [8–10]), but in such a way that the total fermion number (or electric charge) does not change. This is because the number of created right- or left-handed antifermions equals the number of left- or right-handed fermions:

$$N_R + N_L = (\#^R_{1/2} - \#^R_{-1/2}) + (\#^L_{-1/2} - \#^L_{1/2}) = 0. \qquad (1)$$

However, it turns out that the difference in the created number of right-handed and left-handed fermions is not identically zero. This means that it is possible to create a net amount of helicity:

$$N_R - N_L = (\#^R_{1/2} - \#^R_{-1/2}) - (\#^L_{-1/2} - \#^L_{1/2}) = [\#_{1/2} - \#_{-1/2}]. \qquad (2)$$

The simplest scenario where this is possible is for a constant magnetic field, say, in the third spatial direction $\vec{B} = (0,0,B)$, together with a pulse of electric field parallel to it, $\vec{E} = (0,0,E(t))$. One can show that, in this situation, the net creation of helicity per unit volume V is given by (see [1] for a proof involving an adiabatic electric pulse)

$$\frac{\Delta(N_R - N_L)}{V} = -\frac{q^2}{2\pi^2}\int_{t_1}^{t_2} dt\, \vec{E}\cdot\vec{B}, \qquad (3)$$

where q is the electric charge of the fermion, and the fermionic field is assumed to start in the vacuum state at early times before the electric field is switched on. Hence, if the integral $\int_{t_1}^{t_2} dt\,\vec{E}\cdot\vec{B}$ is different from zero, particles with different helicities are created in different amounts. In contrast, $N_R + N_L$ remains strictly constant. For an arbitrary electromagnetic background, the previous result generalizes to

$$\Delta(N_R - N_L) = -\frac{q^2}{2\pi^2}\int_{t_1}^{t_2} dt \int d^3x\, \vec{E}\cdot\vec{B}. \qquad (4)$$

The key point is that Equation (4) is equivalent to the quantum-mechanical symmetry breaking of the fermion chiral symmetry: $\psi \to \psi' = e^{-i\epsilon\gamma^5}\psi$, as expressed in the anomalous non-conservation of the current [6,7].

$$\partial_\mu \langle j_5^\mu \rangle = -\frac{q^2 \hbar}{8\pi^2} F_{\mu\nu} {}^*F^{\mu\nu} \tag{5}$$

where $F_{\mu\nu}$ is the electromagnetic field strength, ${}^*F_{\mu\nu} \equiv \frac{1}{2}\epsilon_{\mu\nu\alpha\beta}F^{\alpha\beta}$ its dual, and the presence of $\hbar$ makes manifest that this is a quantum effect. Equations (5) and (4) are connected by the standard relation between a current and the charge associated with it: $\int d^3x \langle j_5^0 \rangle = \hbar(N_R - N_L)$.

The discovery of the chiral anomaly of Equation (5) was not arrived at by computing the number of fermions created, but rather by directly computing the quantity $\partial_\mu \langle j_5^\mu \rangle$. In that calculation, the anomaly arises from the renormalization subtractions needed to calculate the expectation value $\langle j_5^\mu \rangle$. The operator j_5^μ is non-linear (quadratic) in the fermion field, and in quantum field theory expectation values of non-linear operators are plagued with ultraviolet divergences. One must use renormalization techniques to extract the physical, finite result. A detailed study shows that renormalization methods that respect the gauge invariance of the electromagnetic background break the fermionic chiral symmetry of the classical theory. The fact that Expression (5) was able to accurately explain the decay ratio of processes that could not be understood otherwise, like the decay of the neutral pion to two photons, was an important milestone in the quantum field theory and the study of anomalies.

A similar anomaly appears when the electromagnetic background is replaced by a gravitational field [11]. In this case, renormalization methods that respect general covariance give rise to a violation of the classical conservation law $\nabla_\mu j_5^\mu = 0$, which becomes

$$\nabla_\mu \langle j_5^\mu \rangle = -\frac{\hbar}{192\pi^2} R_{\mu\nu\alpha\beta} {}^*R^{\mu\nu\alpha\beta} \tag{6}$$

where $R_{\mu\nu\alpha\beta}$ is the Riemann tensor and ${}^*R^{\mu\nu\alpha\beta}$ its dual, and ∇_μ is the covariant derivative. For gravitational fields for which a particle interpretation is available at early and late times, this chiral anomaly also manifests in the net helicity contained in the fermionic particles created during the evolution:

$$\Delta(N_R - N_L) = -\frac{1}{192\pi^2} \int_{t_1}^{t_2} \int_\Sigma d^4x \sqrt{-g} \, R_{\mu\nu\alpha\beta} {}^*R^{\mu\nu\alpha\beta} . \tag{7}$$

Typical configurations where this integral is non-zero are the gravitational collapse of a neutron star, or the merger of two compact objects as the ones recently observed by the LIGO-Virgo collaboration [12,13].

In contrast to the anomaly of Equation (5) induced by an electromagnetic background, the chiral anomaly induced by gravity affects every sort of massless spin-1/2 fields, either charged or neutral. This is a consequence of the universal character of gravity, encoded in the equivalence principle, that guarantees that, if Equation (6) is valid for a type of massless spin-1/2 field, it must also be valid for any other type.

There is no reason to believe, however, that these anomalies are specific to spin 1/2 fermions, and one could in principle expect that a similar effect will arise for other types of fields that classically admit chiral-type symmetries. This is the case of photons. One then expects that photons propagating in the presence of a gravitational field will not preserve their net helicity, or, in the language of particles, that the gravitational field will created photons with different helicities in unequal amounts, in the same way it happens for fermions (one also expects a similar effect for gravitons). For photons, the analog of the classical chiral symmetry of fermions is given by electric-magnetic duality rotations [14]

$$\begin{aligned} \vec{E}' &= \cos\theta \, \vec{E} + \sin\theta \, \vec{B} \\ \vec{B}' &= \cos\theta \, \vec{B} - \sin\theta \, \vec{E} . \end{aligned} \tag{8}$$

As first proved in [15], these transformations leave the action of electrodynamics invariant if sources (charges and currents) are not present, and the associated Noether charge is precisely the difference between the intensity of the right- and left-handed circularly polarized electromagnetic waves, i.e., the net helicity. This symmetry is exact in the classical theory even in the presence of an arbitrary gravitational background, as pointed out some years later in [16]. However, in exact analogy with the fermionic case, quantum effects can break this symmetry of the action and induce an anomaly [17,18]. In the language of particles, this would imply that the difference in the number of photons with helicities $h = \pm 1$, $N_R - N_L$, is not necessarily conserved in curved spacetimes. The analogy with the fermionic case also suggests that the current j_D^μ associated with the symmetry under duality rotations of the classical theory fails to be conserved quantum mechanically, with a non-conservation law of the type

$$\nabla_\mu \langle j_D^\mu \rangle = \alpha \, \hbar R_{\mu\nu\alpha\beta} {}^* R^{\mu\nu\alpha\beta} . \tag{9}$$

where α is a numerical coefficient to be determined. In a recent work [18,19], we have proved that this is in fact the case, and obtained that α is different from zero and given by $\alpha = -\frac{1}{96\pi^2}$. In this paper, we will provide a general overview of these results from a different perspective, and with more emphasis on conceptual aspects.

2. Electric-Magnetic Duality Rotations and Self- and Anti Self-Dual Fields

To study electric-magnetic rotations of Equation (8), it is more convenient to change variables to the self- and anti-self-dual components of the electromagnetic field, defined by $\vec{H}_\pm \equiv \frac{1}{\sqrt{2}} (\vec{E} \pm i \vec{B})$, since for them the transformation of Equation (8) takes a diagonal form:

$$\vec{H}'_\pm = e^{\mp i\theta} \vec{H}_\pm . \tag{10}$$

A discrete duality transformation $\star \vec{E} = \vec{B}$, $\star \vec{B} = -\vec{E}$ corresponds to $\theta = \pi/2$. Then, the operator $i\star$ produces $i \star \vec{H}_\pm = \pm \vec{H}_\pm$. It is for this reason that $\vec{H}_+$ and $\vec{H}_-$ are called the self- and anti-self-dual components of the electromagnetic field, respectively.

There are other aspects that support the convenience of these variables. For instance, under a Lorentz transformation, the components of $\vec{E}$ and $\vec{B}$ mix with each other. Indeed, under an infinitesimal Lorentz transformation of rapidity $\vec{\eta}$, the electric and magnetic fields transform as $\vec{E}' = \vec{E} - \vec{\eta} \wedge \vec{B}$, $\vec{B}' = \vec{B} + \vec{\eta} \wedge \vec{E}$. (We recall that the rapidity $\vec{\eta}$ completely characterizes a Lorentz boost: its modulus contains the information of the Lorentz factor γ, via $\cosh |\vec{\eta}| = \gamma$, and its direction indicates the direction of the boost). However, when $\vec{E}$ and $\vec{B}$ are combined into $\vec{H}_\pm$, it is easy to see that the components of $\vec{H}_+$ and $\vec{H}_-$ no longer mix:

$$\vec{H}'_\pm = \vec{H}_\pm \pm i\vec{\eta} \wedge \vec{H}_\pm . \tag{11}$$

Note also that, under an ordinary infinitesimal (counterclockwise) rotation of angle $\alpha > 0$ around the direction of a unit vector $\vec{n}$, the complex vectors $\vec{H}_\pm$ transform as $\vec{H}'_\pm = \vec{H}_\pm + \alpha \vec{n} \wedge \vec{H}_\pm$. Hence, a boost corresponds to a rotation of an imaginary angle. These are the transformation rules associated with the two irreducible representations of the Lorentz group for fields of spin $s = 1$. In the standard terminology [20,21], they correspond to the $(0,1)$ representation for $\vec{H}_+$, and the $(1,0)$ one for $\vec{H}_-$. More generally, for any element of the restricted Lorentz group $SO^+(1,3)$ (rotations + boots), the above complex fields transform as

$$\vec{H}'_\pm = e^{-i(\alpha\vec{n} \pm i\vec{\eta}) \cdot \vec{J}} \vec{H}_\pm \tag{12}$$

where $\vec{J}$ are the infinitesimal generators of the group of rotations. The $\pm$ sign in the above equation distinguishes the two inequivalent (three-dimensional) representations of the Lorentz group. They are, however, equivalent under the subgroup of rotations. This makes transparent the fact that electrodynamics describes fields of spin $s = 1$, something that is more obscure when working with $\vec{E}$ and $\vec{B}$, the field strength $F_{\mu\nu}$, or even the vector potential A_μ.

Another useful aspect of self- and anti-self-dual variables concerns the equations of motion. The source-free Maxwell equations

$$\vec{\nabla} \cdot \vec{E} = 0, \qquad \vec{\nabla} \cdot \vec{B} = 0$$
$$\vec{\nabla} \times \vec{E} = -\partial_t \vec{B}, \qquad \vec{\nabla} \times \vec{B} = \partial_t \vec{E} \qquad (13)$$

when written in terms of $\vec{H}_\pm$, take the form

$$\vec{\nabla} \cdot \vec{H}_\pm = 0, \qquad \vec{\nabla} \times \vec{H}_\pm = \pm i \partial_t \vec{H}_\pm. \qquad (14)$$

Notice that, in contrast to $\vec{E}$ and $\vec{B}$, the self- and anti-self-dual fields are not coupled by the dynamics. The general solution to these field equations is a linear combination of positive and negative frequency plane waves

$$\vec{H}_\pm(t, \vec{x}) = \int \frac{d^3k}{(2\pi)^3} \left[h_\pm(\vec{k}) e^{-i(kt - \vec{k} \cdot \vec{x})} + h_\mp^*(\vec{k}) e^{i(kt - \vec{k} \cdot \vec{x})} \right] \hat{e}_\pm(\vec{k}) \qquad (15)$$

where $k = |\vec{k}|$ and $h_\pm(\vec{k})$ are complex numbers that quantify the wave amplitudes. The polarization vectors are given by $\hat{e}_\pm(\vec{k}) = \frac{1}{\sqrt{2}}(\hat{e}_1(\vec{k}) \pm i \hat{e}_2(\vec{k}))$ where $\hat{e}_1$ and $\hat{e}_2$ are any two real, space-like unit vectors transverse to $\hat{k}$ (we choose their orientation such that $\hat{e}_1 \times \hat{e}_2 = +\hat{k}$). Positive-frequency Fourier modes $h_\pm(\vec{k}) e^{-i(kt - \vec{k} \cdot \vec{x})} \hat{e}_\pm(\vec{k})$ describe waves with helicity $h = 1$ for self-dual fields, and with negative helicity $h = -1$ for anti-self-dual fields. This is also in agreement with the general fact that a massless field associated with the Lorentz representation $(0, j)$ describes particles with helicity $+j$, while a $(j, 0)$-field describes particles with helicity $-j$ [20]. Compared with massless fermions, $\vec{H}_+$ is the analog of a right-handed Weyl spinor, which transforms under the $(0, 1/2)$ Lorentz representation, and $\vec{H}_-$ is the analog of a left-handed Weyl spinor.

The constraints $\vec{\nabla} \cdot \vec{H}_\pm = 0$ can be used to introduce the potentials $\vec{A}_\pm$, as follows:

$$\vec{H}_\pm = \pm i \vec{\nabla} \times \vec{A}_\pm. \qquad (16)$$

Maxwell equations then reduce to first-order differential equations for the potentials:

$$\pm i \vec{\nabla} \times \vec{A}_\pm = -\partial_t \vec{A}_\pm + \vec{\nabla} A_\pm^0. \qquad (17)$$

Both sets of equations, for the fields of Equation (14) and for the potentials of Equation (17), can be written more compactly as follows (the equations for $\vec{H}_-$ and $\vec{A}_-$ are obtained by complex conjugation)

$$\alpha_I^{ab} \partial_a H_+^I = 0, \qquad \bar{\alpha}_I^{ab} \partial_a A_{+b} = 0. \qquad (18)$$

The numerical constants α_I^{ab} are three 4×4 matrices, for $I = 1, 2, 3$, and the bar over α_I^{ab} indicates complex conjugation. The components of these matrices in an inertial frame are

$$\alpha_1^{ab} = \begin{pmatrix} 0 & -1 & 0 & 0 \\ 1 & 0 & 0 & 0 \\ 0 & 0 & 0 & i \\ 0 & 0 & -i & 0 \end{pmatrix} \quad \alpha_2^{ab} = \begin{pmatrix} 0 & 0 & -1 & 0 \\ 0 & 0 & 0 & -i \\ 1 & 0 & 0 & 0 \\ 0 & i & 0 & 0 \end{pmatrix} \quad \alpha_3^{ab} = \begin{pmatrix} 0 & 0 & 0 & -1 \\ 0 & 0 & i & 0 \\ 0 & -i & 0 & 0 \\ 1 & 0 & 0 & 0 \end{pmatrix}. \qquad (19)$$

It is trivial to check by direct substitution that Equation (18) is equivalent to Equations (14) and (17), respectively. These anti-symmetric matrices are Lorentz invariant symbols. They are self-dual ($i \star \alpha_I^{ab} = \alpha_I^{ab}$), and the conjugate matrices are anti-self-dual ($i \star \bar{\alpha}_I^{ab} = -\bar{\alpha}_I^{ab}$).

The two sets of equations in Equation (18) were shown in [19] to contain the same information. One can thus formulate source-free Maxwell theory entirely in terms of complex potentials.

3. Noether Symmetry and Conserved Charge

In this section, we show that electric-magnetic rotations of Equation (8) are a symmetry of the classical theory, and obtain an expression for the associated conserved charge. This can be more easily done by working in Hamiltonian formalism. The phase space of electrodynamics is usually parametrized by the pair of fields $(\vec{A}(\vec{x}), \vec{E}(\vec{x}))$, with $\vec{B} = \vec{\nabla} \times \vec{A}$. The Hamiltonian of the theory is easily obtained by the Legendre transform from the Lagrangian, and it reads

$$H = \frac{1}{2} \int d^3x \left[\vec{E}^2 + (\vec{\nabla} \times \vec{A})^2 - A_0 (\vec{\nabla} \cdot \vec{E}) \right]. \tag{20}$$

In this expression, $A_0(\vec{x})$ is regarded as a Lagrangian multiplier that enforces the Gauss law constraint $\vec{\nabla} \cdot \vec{E} = 0$. The phase space is equipped with a Poisson structure given by $\{A_i(\vec{x}), E^j(\vec{x}')\} = \delta_i^j \delta^{(3)}(\vec{x} - \vec{x}')$, which induces a natural symplectic product $\Omega[(\vec{A}_1, \vec{E}_1), (\vec{A}_2, \vec{E}_2)] = -\frac{1}{2} \int d^3x \left[\vec{E}_1 \cdot \vec{A}_2 - \vec{E}_2 \cdot \vec{A}_1 \right]$.

From the form of the electric-magnetic rotations of Equation (8), we see that the infinitesimal transformation of the canonical variables reads

$$\delta \vec{A} = \vec{Z}, \qquad \delta \vec{E} = \vec{\nabla} \times \vec{A} \tag{21}$$

where $\vec{Z}$ is defined by $\vec{E} =: -\vec{\nabla} \times \vec{Z}$; therefore, it can be understood as an "electric potential" (note that in the source-free theory $\vec{Z}$ can be always defined, since $\vec{\nabla} \cdot \vec{E} = 0$).

Now, the generator of the transformation of Equation (21) can be determined by

$$Q_D = \Omega[(\vec{A}, \vec{E}), (\delta \vec{A}, \delta \vec{E})] = -\frac{1}{2} \int d^3x \left[\vec{E} \cdot \delta \vec{A} - \vec{A} \cdot \delta \vec{E} \right] = \frac{1}{2} \int d^3x \left[\vec{A} \cdot \vec{B} - \vec{Z} \cdot \vec{E} \right]. \tag{22}$$

Q_D is gauge invariant, and one can easily check that it generates the correct transformation by computing Poisson brackets

$$\begin{aligned} \delta \vec{B} &= \{\vec{B}, Q_D\} = \{\nabla \times \vec{A}, Q_D\} = -\vec{E} \\ \delta \vec{E} &= \{\vec{E}, Q_D\} = \vec{B}. \end{aligned} \tag{23}$$

It is also straightforward to check that $\delta H = \{H, Q\} = 0$. Therefore, the canonical transformation generated by Q_D, i.e., the electric-magnetic duality transformation of Equation (21), is a symmetry of the source-free Maxwell theory, and Q_D is a constant of motion.

Taking into account the form of the generic solutions, Equation (15), to the field equations, the conserved charge reads

$$Q_D = \int \frac{d^3k}{(2\pi)^3 k} \left[|h_+(\vec{k})|^2 - |h_-(\vec{k})|^2 \right]. \tag{24}$$

This expression makes it clear that Q_D is proportional to the difference in the intensity of the self- and anti-self-dual parts of field or, equivalently, the difference between the right and left circularly polarized components. In the quantum theory, $Q_D/\hbar$ measures the difference in the number of photons with helicities $h = +1$ and $h = -1$. For this reason, we recognize Q_D as the V-Stokes parameter that describes the polarization state of the electromagnetic radiation.

Although we have restricted here to Minkowski spacetime, the argument generalizes to situations in which a gravitational field is present [16]. A generally covariant proof in curved spacetimes in the Lagrangian formalism is given in [19], where the associated Noether current was obtained:

$$j_D^\mu = \frac{1}{2} \left[A_\nu {}^* F^{\mu\nu} - Z_\nu F^{\mu\nu} \right]. \tag{25}$$

4. Analogy with Dirac Fermions and the Quantum Anomaly

The goal of this section is to compute the vacuum expectation value of the current j_D^μ associated with the symmetry under electric-magnetic rotations, and to use the result to evaluate whether these transformations are also a symmetry of the quantum theory. A convenient strategy to achieve this is to realize that, in the absence of electric charges and currents, Maxwell's theory can be formally written as a (bosonic) spin 1 version of the Dirac theory for a real spin 1/2 field. The convenience of writing the theory in this form is that it allows one to take advantage of numerous and powerful tools developed to compute the chiral anomaly for fermions. Hence, we will start in Section 4.1 by summarizing the theory of massless spin 1/2 fermions and the calculation of the fermionic chiral anomaly, and we will come back to the electromagnetic case in Section 4.2.

4.1. Fermions in Curved Spacetime

To better motivate the analogy between electric-magnetic rotations and chiral rotations of fermions, it is convenient to write the Dirac field in terms of two Weyl spinors ψ_L and ψ_R as follows (see for instance [1,22]):

$$\psi \equiv \begin{pmatrix} \psi_L \\ \psi_R \end{pmatrix}, \quad \bar\psi \equiv \psi^\dagger \beta = (\psi_L^\dagger, \psi_R^\dagger) \tag{26}$$

where β is the matrix

$$\beta \equiv \begin{pmatrix} 0 & I \\ I & 0 \end{pmatrix}. \tag{27}$$

The spinor ψ_L transforms according to the $(1/2, 0)$ representation of the Lorentz algebra, while the spinor ψ_R transforms with the $(0, 1/2)$ representation. The Dirac equation

$$i\gamma^\mu \partial_\mu \psi = m\psi \tag{28}$$

takes the form

$$i \begin{pmatrix} 0 & \sigma^\mu \\ \bar\sigma^\mu & 0 \end{pmatrix} \begin{pmatrix} \psi_L \\ \psi_R \end{pmatrix} = m \begin{pmatrix} \psi_L \\ \psi_R \end{pmatrix} \tag{29}$$

where $\sigma^\mu = (I, \vec\sigma)$ and $\vec\sigma$ are the Pauli matrices. Numerically β agrees with the Dirac matrix γ^0, and it is for this reason that the two matrices are commonly identified (although they have a different index structure; see e.g., [22]). For massless fermions, the theory is invariant under the chiral transformations $\psi \to \psi' = e^{i\theta\gamma_5}\psi$, with $\gamma_5 = \frac{i}{4!}\epsilon_{\alpha\beta\gamma\delta}\gamma^\alpha\gamma^\beta\gamma^\gamma\gamma^\delta$

$$\gamma_5 = \begin{pmatrix} -I & 0 \\ 0 & I \end{pmatrix}. \tag{30}$$

Therefore,

$$\psi = \begin{pmatrix} \psi_L \\ \psi_R \end{pmatrix} \to \psi' = e^{i\gamma_5\theta}\psi = \begin{pmatrix} e^{-i\theta}\psi_L \\ e^{i\theta}\psi_R \end{pmatrix}. \tag{31}$$

Noether's theorem associates with this symmetry transformation the chiral current $j_5^\mu = \bar\psi\gamma^\mu\gamma_5\psi$. The spatial integral of its time-component is the charge

$$Q_5 = \int d^3x (\psi_R^\dagger \psi_R - \psi_L^\dagger \psi_L), \tag{32}$$

and it is classically conserved.

This charge counts the difference in the number of positive and negative helicity states, in close analogy to the dual charge of Equation (24) for the electromagnetic case. As we mentioned in the introduction, this quantity is a constant of motion in the quantum theory in Minkowski space, but this is not necessarily true in the presence of a gravitational background, as we now explain in more detail.

In the presence of gravity, the Dirac equation for a massless spin 1/2 fields takes the form (see, for instance, [23])

$$i\gamma^\mu(x)\nabla_\mu \psi(x) = 0 \tag{33}$$

where $\gamma^\mu(x) = e^\mu_a(x)\gamma^a$ are the Dirac gamma matrices in curved space, $e^\mu_a(x)$ is a Vierbein or orthonormal tetrad in terms of which the curved metric $g_{\mu\nu}$ is related to the Minkowski metric η_{ab} by $g_{\mu\nu} e^\mu_a e^\nu_b = \eta_{ab}$, while γ^a are the Minkowskian gamma matrices (that satisfy $\{\gamma^a, \gamma^b\} = 2\eta^{ab}$). ∇_μ is the covariant derivate acting on spin 1/2 fields:

$$\nabla_\mu \psi = (\partial_\mu + i\omega_{\mu ab}\Sigma^{ab})\psi \tag{34}$$

where $\Sigma^{ab} = -\frac{1}{8}[\gamma^a, \gamma^b]$ are the generators of the $(1/2, 0) \oplus (0, 1/2)$ representation of the Lorentz group, and w_μ is the standard spin connexion, defined in terms of the Vierbein and the Christoffel symbols $\Gamma^\alpha_{\mu\beta}$ by $(w_\mu)^a{}_b = e^a_\alpha \partial_\mu e^\alpha_b + e^a_\alpha e^\beta_b \Gamma^\alpha_{\mu\beta}$.

The axial symmetry is maintained at the classical level, or in other words, the conservation law $\nabla_\mu j^\mu_A = 0$ holds for any solution of the equations of motion. Quantum mechanically, to check whether the symmetry is maintained one needs to evaluate the vacuum expectation value of the operator $\nabla_\mu j^\mu_A$. The result, originally computed in [11], is given by

$$\langle \nabla_\mu j^\mu_A \rangle = \frac{2i\hbar}{(4\pi)^2} tr[\gamma_5 E_2(x)] \tag{35}$$

where $E_2(x)$ is the second DeWitt coefficient (see the appendix for a sketch of the derivation, and [23] for a pedagogical calculation using different renormalization methods). In short, the DeWitt coefficients are local functions constructed from curvature tensors that encode the information of the short distance behavior $(x' \to x)$ of the solution $K(\tau, x, x')$ of a heat-type equation associated with the Dirac operator $D \equiv i\gamma^\mu \nabla_\mu$ (for this reason, this function K is called the Heat-Kernel):

$$i\partial_\tau K(\tau, x, x') = D^2 K(\tau, x, x') . \tag{36}$$

The asymptotic form of $K(\tau, x, x)$ as $\tau \to 0$ defines the $E_n(x)$ coefficients by

$$K(\tau, x, x) \sim \frac{-i}{(4\pi\tau)^2} \sum_{n=0}^\infty (i\tau)^n E_n(x) . \tag{37}$$

$E(x)$ are local quantities encoding analytical information of the Klein–Gordon operator D^2 in Equation (36)

$$D^2 \psi = (g^{\mu\nu} \nabla_\mu \nabla_\nu + \mathcal{Q}(x))\psi = 0 \tag{38}$$

and are determined by the geometry of the spacetime background. The result for the $E_2(x)$ is [23]

$$\begin{aligned} E_2(x) &= \left[-\frac{1}{30}\Box R + \frac{1}{72}R^2 - \frac{1}{180}R_{\mu\nu}R^{\mu\nu} + \frac{1}{180}R_{\alpha\beta\mu\nu}R^{\alpha\beta\mu\nu}\right]\mathbb{I} \\ &+ \frac{1}{12}W_{\mu\nu}W^{\mu\nu} + \frac{1}{2}\mathcal{Q}^2 - \frac{1}{6}R\mathcal{Q} + \frac{1}{6}\Box \mathcal{Q} \end{aligned} \tag{39}$$

where $W_{\mu\nu}$ is defined by $W_{\mu\nu}\psi = [\nabla_\mu, \nabla_\nu]\psi$, and

$$Q = \frac{1}{4}R, \quad W_{\mu\nu} = -iR_{\mu\nu\alpha\beta}e^\alpha_a e^\beta_b \Sigma^{ab}. \tag{40}$$

The non-trivial contribution to the axial anomaly comes entirely from the $W_{\mu\nu}W^{\mu\nu}$ term and produces

$$\begin{aligned} \langle \nabla_\mu j^\mu_A \rangle &= \frac{2i\hbar}{(4\pi)^2} tr[\gamma_5 E_2(x)] = -\frac{2i\hbar}{(4\pi)^2}\frac{1}{12} R_{\mu\nu ab} R^{\mu\nu cd} tr[\gamma_5 \Sigma^{ab}\Sigma_{cd}] \\ &= \frac{\hbar}{192\pi^2} R_{\mu\nu\lambda\sigma}{}^* R^{\mu\nu\lambda\sigma}. \end{aligned} \tag{41}$$

If, in addition to the gravitational background, the fermion field propagates also on an electromagnetic background, there is another contribution to the anomaly (this one is proportional to the square of the electric charge q of the fermion). The extra contributions to $W_{\mu\nu}$ and Q are $W_{\mu\nu} = iqF_{\mu\nu}$ and $Q = 2qF_{\mu\nu}\Sigma^{\mu\nu}$, and the expression for $\langle \nabla_\mu j^\mu_A \rangle$ becomes

$$\langle \nabla_\mu j^\mu_A \rangle = \frac{\hbar}{192\pi^2} R_{\mu\nu\lambda\sigma}{}^* R^{\mu\nu\lambda\sigma} - \frac{\hbar q^2}{8\pi^2} F_{\mu\nu}{}^* F^{\mu\nu}. \tag{42}$$

To finish this section, recall that there is another type of spin 1/2 fermions known as Majorana spinors. They are the "real" versions of Dirac's spinors. Mathematically, while for Dirac massless fermions the two Weyl spinors ψ_L and ψ_R in Equation (26) are independent of each other, this is not true for Majorana spinors, for which there is an extra condition $\psi_R = i\sigma^2 \psi^*_L$ [1]. Furthermore, the Lagrangian density for Majorana spinors carries an additional normalization factor 1/2 compared to Dirac's Lagrangian. Since Majorana spinors do not carry an electric charge ($q = 0$), the presence of an electromagnetic background does not induce any anomaly, and the coefficient in the gravitational sector of the anomaly is half of the value obtained for a Dirac fermion.

4.2. Electrodynamics in Curved Spacetime

Consider Maxwell theory in the absence of electric charges and currents. This theory can be described by a classical action that is formally analog to the action of a Majorana 4-spinor. Rather than proving from scratch that the familiar Maxwell action can be re-written in the form just mentioned (see [19]), we will simply postulate the new action and show then that it reproduces the correct equations of motion. Consider then the following action in terms of self-dual and anti self-dual variables:

$$S[A^+, A^-] = -\frac{1}{4}\int d^4x \sqrt{-g}\, \bar{\Psi}\, i\beta^\mu \nabla_\mu \Psi \tag{43}$$

where

$$\Psi = \begin{pmatrix} A^+ \\ H_+ \\ A^- \\ H_- \end{pmatrix}, \quad \bar\Psi = (A^+, H_+, A^-, H_-), \quad \beta^\mu = i\begin{pmatrix} 0 & 0 & 0 & \bar\alpha^\mu \\ 0 & 0 & -\alpha^\mu & 0 \\ 0 & \alpha^\mu & 0 & 0 \\ -\bar\alpha^\mu & 0 & 0 & 0 \end{pmatrix}. \tag{44}$$

Note that Ψ is formally analog to a Majorana 4-spinor rather than a Dirac one, since its lower two components are complex conjugate from the upper ones. Therefore, the action of Equation (43) is the analog of Majorana's action. The independent variables in this action are the potentials $A^\mu_\pm$, and the fields $\bar H_\pm$ are understood as shorthands for their expressions in terms of the potentials (see Section 2). Note also that Equation (43) is a first-order action (i.e., first-order in time derivatives), while the standard Maxwell's action is second order. ∇_μ in Equation (43) is the covariant derivative acting on the field Ψ, given by

$$\nabla_\mu \Psi = (\partial_\mu + i\omega_{\mu ab} M^{ab})\Psi \tag{45}$$

and M^{ab} is

$$M^{ab} = \begin{pmatrix} \Sigma^{ab} & 0 & 0 & 0 \\ 0 & {}^+\Sigma^{ab} & 0 & 0 \\ 0 & 0 & \Sigma^{ab} & 0 \\ 0 & 0 & 0 & {}^-\Sigma^{ab} \end{pmatrix} \tag{46}$$

where $\Sigma^{\sigma\rho}{}_{\alpha\beta} = \delta^\rho_\alpha \delta^\sigma_\beta - \delta^\rho_\beta \delta^\sigma_\alpha$ is the generator of the $(1/2, 1/2)$ representation of the Lorentz group, while ${}^+\Sigma^{\sigma\rho}_{IJ}$ and ${}^-\Sigma^{\sigma\rho}_{IJ}$ are the generators of the $(0,1) \oplus (0,0)$ and $(1,0) \oplus (0,0)$ representations, respectively.

Using some algebraic properties of the matrices α (see [19] for more details), it is not difficult to find that β^μ satisfies the Clifford algebra

$$\{\beta^\mu, \beta^\nu\} = 2g^{\mu\nu}\mathbb{I}. \tag{47}$$

It can also be checked that $\nabla_\nu \beta^\mu(x) = 0$. These matrices can then be thought of as the spin 1 counterpart of the Dirac γ^μ matrices. Furthermore, one can also introduce the "chiral" β_5 matrix in a similar way:

$$\beta_5 \equiv \frac{i}{4!} \epsilon_{\alpha\beta\gamma\delta} \beta^\alpha \beta^\beta \beta^\gamma \beta^\delta = \begin{pmatrix} -\mathbb{I} & 0 & 0 & 0 \\ 0 & -\mathbb{I} & 0 & 0 \\ 0 & 0 & \mathbb{I} & 0 \\ 0 & 0 & 0 & \mathbb{I} \end{pmatrix}, \tag{48}$$

satisfying properties analogous to the Dirac case:

$$\{\beta^\mu, \beta_5\} = 0, \qquad \beta_5^2 = \mathbb{I}. \tag{49}$$

Further details and properties of these matrices can be studied in [19].

Although the basic variables in the action are the potentials $A^\pm_\mu$, at the practical level one can work by considering Ψ and $\bar{\Psi}$ as independent fields. Note that this is the same as one does when working with Majorana spinors. The equations of motion take the form

$$\frac{\delta S}{\delta \bar{\Psi}} = 0 \quad \longrightarrow \quad i\beta^\mu \nabla_\mu \Psi = 0. \tag{50}$$

They contain four equations, one for each of the four components of Ψ. The upper two are the equations $\tilde{\alpha}^{\mu\nu}_I \nabla_\mu A^+_\nu = 0$ and $\alpha^{\mu\nu}_I \nabla_\mu H^I_+ = 0$. The lower two are complex conjugated equations. Since these equations are precisely Maxwell's equations written in self- and anti-self-dual variables, this proves that the action of Equation (43) describes the correct theory.

Now we study how the classical electric-magnetic symmetry and its related conservation law arise in this formalism. By means of the chiral matrix β_5, an electric-magnetic duality rotation can be written in the following form, manifestly analog to a chiral transformation for Dirac fields:

$$\Psi \to e^{i\theta\beta_5}\Psi, \qquad \bar{\Psi} \to \bar{\Psi} e^{i\theta\beta_5}. \tag{51}$$

Recalling the explicit form of β_5 in Equation (48), one infers that the upper two components of Ψ, namely (A_+, H_+), encode the self-dual, or positive chirality sector of the theory, while the lower two components (A_-, H_-) describe the anti-self-dual or the negative chiral sector. The Lagrangian density in Equation (43) remains manifestly invariant under these rotations, and in the language of Ψ the Noether current reads

$$j_D^\mu = \frac{1}{4}\bar\Psi \beta^\mu \beta_5 \Psi \,. \tag{52}$$

The corresponding Noether charge yields

$$Q_D = \int_{\Sigma_t} d\Sigma_\mu\, j_D^\mu = \frac{1}{4}\int_{\Sigma_t} d\Sigma_3\, \bar\Psi \beta^0 \beta_5 \Psi \tag{53}$$

where $d\Sigma_3$ is the volume element of a space-like Cauchy hypersurface Σ_t. This formula for Q_D is in full agreement to that calculated in previous sections (see Equation (22)), generalized to curved spacetimes.

The calculation of the vacuum expectation value $\langle \nabla_\mu j_D^\mu \rangle$ in the quantum theory follows exactly the same steps shown above for fermions. Namely, $\langle \nabla_\mu j_D^\mu \rangle$ is given again [18,19] in terms of the second DeWitt coefficient $E_2(x)$ by

$$\langle \nabla_\mu j_D^\mu \rangle = -i\frac{\hbar}{32\pi^2}\, tr[\beta_5 E_2]\,, \tag{54}$$

where $E_2(x)$ is now obtained from the heat kernel K associated with the *Maxwell operator* $D = i\beta^\mu \nabla_\mu$, rather than the Dirac operator $i\gamma^\mu \nabla_\mu$. The DeWitt coefficient is still given by Equation (39), but now Equation (40) needs to be replaced by

$$\mathcal{Q}\Psi \equiv \frac{1}{2}\beta^{[\alpha}\beta^{\mu]} W_{\alpha\mu} \Psi \tag{55}$$

and

$$W_{\alpha\mu}\Psi \equiv [\nabla_\alpha, \nabla_\mu]\Psi = \frac{1}{2} R_{\alpha\mu\sigma\rho} M^{\sigma\rho} \Psi \,. \tag{56}$$

With this, Equation (54) becomes

$$\langle \nabla_\mu j_D^\mu \rangle_{\rm ren} = -\frac{\hbar}{96\pi^2} R_{\alpha\beta\mu\nu}\,{}^\star R^{\alpha\beta\mu\nu} \,. \tag{57}$$

This result reveals that quantum fluctuations spoil the conservation of the axial current j_D^μ and break the classical symmetry under electric-magnetic (or chiral) transformations, if the spacetime curvature is such that the curvature invariant $R_{\alpha\beta\mu\nu}\,{}^\star R^{\alpha\beta\mu\nu}$ is different from zero.

5. Discussion

The result shown in Equation (57) implies that the classical Noether charge Q_D is not necessarily conserved in the quantum theory, and its change between two instants t_1 and t_2 can be written as

$$\Delta \langle Q_D \rangle = -\frac{\hbar}{96\pi^2}\int_{t_1}^{t_2}\int_\Sigma d^4 x\sqrt{-g}\, R_{\alpha\beta\mu\nu}\,{}^\star R^{\alpha\beta\mu\nu} = -\frac{\hbar}{6\pi^2}\int_{t_1}^{t_2} dt \int_\Sigma d^3 x \sqrt{-g}\, E_{\mu\nu} B^{\mu\nu} \tag{58}$$

where in the last equality we have written $R_{\alpha\beta\mu\nu}\,{}^\star R^{\alpha\beta\mu\nu}$ in terms of the electric $E_{\mu\nu}$ and magnetic $B_{\mu\nu}$ parts of the Weyl curvature tensor. Note the close analogy with the chiral spin $1/2$ anomaly shown in Equation (4). This result implies that the polarization state of the quantum electromagnetic field can change in time, even in the complete absence of electromagnetic sources, due to the influence of gravitational dynamics and quantum electromagnetic effects (notice the presence of $\hbar$). In this precise sense, one can think about the spacetime as an optically active medium.

Since $\Delta \langle Q_D \rangle$ is proportional to $\hbar$, one could expect the net effect of the anomaly to be small. However, recall that $\Delta \langle Q_D \rangle = \hbar(N_R - N_L)$. Thus, the net number $N_R - N_L$ is only given by the (dimensionless) geometric integral on the RHS of Equation (58). A sufficiently strong gravitational background could lead to a significant effect. It is also important to remark that Expression (58) accounts for the net helicity created out of an initial *vacuum* state—we call this *spontaneous* creation of helicity. However, it is well-known in the study of particle creation by gravitational fields that the spontaneous creation effect for bosons always comes together with the *stimulated* counterpart, if the initial state is not the vacuum but rather contains quanta on it (see [2,24,25]). The stimulated effect is enhanced by the number of initial quanta. For the same reason, the value of $\Delta \langle Q_D \rangle$ is expected to be enhanced if the initial state of radiation is not the vacuum but rather an excited state, as for instance a coherent state which describes accurately the radiation emitted by, say, an astrophysical object. However, remember that the average number of photons in such a coherent state is macroscopic, so it can lead to detectable effects. Therefore, it is conceivable that the change in the polarization of electromagnetic radiation crossing a region of strong gravitational field, produced for instance by the merger of two compact objets, takes macroscopic values. The computation of the exact value of the RHS of Equation (58) in such a situation requires the use of numerical relativity techniques, and this will be the focus of a future project.

Finally, we want to mention that the experimental investigation of this anomaly could be relevant in other areas of physics, as in condensed matter physics [26], non-linear optics [27], or analogue gravity in general. For instance, metamaterials can be designed to manifest properties that are difficult to find in nature [28]. In this case, the medium, and not a distribution of mass-energy, can originate effective geometries [27]. They thus may mimic a curved spacetime with optimal values of Equation (58) and could serve to test the photon right–left asymmetry originating from the electric-magnetic quantum anomaly.

Author Contributions: The authors contributed equally in writing this article. All authors read and approved the final manuscript.

Funding: This research was funded with Grants. No. FIS2014-57387-C3-1-P, No. FIS2017-84440-C2-1-P, No. FIS2017-91161-EXP, No. SEV-2014-0398, and No. SEJI/2017/042 (Generalitat Valenciana), COST action CA15117 (CANTATA), supported by COST (European Cooperation in Science and Technology), NSF CAREER Grant No. PHY-1552603, ERC Consolidator Grant No. MaGRaTh-646597.

Acknowledgments: We thank A. Ashtekar, P. Beltran, E. Bianchi, A. Ferreiro, S. Pla, and J. Pullin for useful discussions.

Conflicts of Interest: The authors declare no conflict of interest.

Appendix A. Some Details Regarding the Calculation of $\nabla_\mu j_A^\mu$ in Curved Spacetimes

In this appendix, we give a sketch of the derivation of Equation (35). The operator of interest, $\nabla_\mu j_A^\mu$, is quadratic in the fermion fields, and thus suffers from ultraviolet (UV) divergences. As a consequence, its vacuum expectation value must incorporate renormalization counterterms in order to cancel out all of them and to provide a finite physically reasonable result:

$$\langle \nabla_\mu j_A^\mu \rangle_{\text{ren}} = \langle \nabla_\mu j_A^\mu \rangle - \langle \nabla_\mu j_A^\mu \rangle_{\text{Ad}(4)}. \tag{A1}$$

Here, $\langle \nabla_\mu j_A^\mu \rangle_{\text{Ad}(4)}$ denotes the (DeWitt–Schwinger) asymptotic expansion up to the fourth adiabatic order [23]. Namely, the renormalization method works by expressing $\langle \nabla_\mu j_A^\mu \rangle$ in terms of the Feymann two-point function $S(x,x') = -i \langle T\Psi(x)\bar\Psi(x') \rangle$ and then replacing $S(x,x')$ with $[S(x,x') - S(x,x')_{\text{Ad}(4)}]$, where $S(x,x')_{\text{Ad}(4)}$ denotes the DeWitt-Schwinger subtractions up to the fourth adiabatic order, and finally taking the limit $x \to x'$.

It is convenient to introduce an auxiliary parameter $s > 0$ in order to regularize spurious infrared divergences in intermediate steps; s will be set to zero at the end of the calculation. This parameter is introduced by replacing the wave equation $D\Psi = 0$ by $(D+s)\Psi = 0$, where $D \equiv i\gamma^\mu \nabla_\mu$. As a result,

$$\begin{aligned}
\nabla_\mu j_A^\mu(x) &= \nabla_\mu [\bar\Psi(x)\gamma^\mu \gamma_5 \Psi(x)] = -i\left[\bar\Psi(x)\overleftarrow{D}\gamma_5 \Psi(x) - \bar\Psi(x)\gamma_5 \overrightarrow{D}\Psi(x)\right] \\
&= \lim_{\substack{s\to 0 \\ x\to x'}} -2is\,\bar\Psi(x)\gamma_5\Psi(x') = \lim_{\substack{s\to 0 \\ x\to x'}} -2is\,\text{Tr}[\gamma_5 \Psi(x)\bar\Psi(x')]
\end{aligned} \tag{A2}$$

where we have used $\{\gamma^\mu, \gamma_5\} = 0$. Picking up an arbitrary vacuum state $|0\rangle$, we have

$$\langle \nabla_\mu j_A^\mu \rangle = \lim_{\substack{s \to 0 \\ x \to x'}} 2s \, \text{Tr}\left[\gamma_5 \, S(x, x', s)\right], \tag{A3}$$

and the renormalized expectation value is given by

$$\langle \nabla_\mu j_A^\mu \rangle_{\text{ren}} = \lim_{\substack{s \to 0 \\ x \to x'}} \frac{1}{2} s \, \text{Tr}\left[\gamma_5 \left(S(x, x', s) - S(x, x', s)_{\text{Ad}(4)}\right)\right]. \tag{A4}$$

Here, $S(x, x', s)$ encodes the information of the vacuum state, and the role of $S(x, x', s)_{\text{Ad}(4)}$ is to remove the ultra-violet divergences—which are the same regardless of the choice of vacuum. It is now useful to write $S(x, x', s)_{\text{Ad}(4)} = [(D_x - s)G(x, x', s)]_{\text{Ad}(4)}$, since it is known that [23]

$$G(x, x', s) \sim \frac{\hbar \Delta^{1/2}(x, x')}{16\pi^2} \sum_{k=0}^\infty E_k(x, x') \int_0^\infty d\tau \, e^{-i\left(\tau s^2 + \frac{\sigma(x,x')}{2\tau}\right)} (i\tau)^{(k-2)}. \tag{A5}$$

In this expression, $\sigma(x, x')$ represents half of the geodesic distance squared between x and x', $\Delta^{1/2}(x, x')$ is the Van Vleck-Morette determinant, and $E_k(x, x')$ are the DeWitt coefficients introduced in the main text ($E_k(x) \equiv \lim_{x' \to x} E_k(x, x')$).

We can safely take now the limit $x = x'$ in which the two points merge. Due to the symmetry of the classical theory, the bare contribution $S(x, x', s)$ in Equation (A4) vanishes for any choice of vacuum state. As a result, $\langle \nabla_\mu j_A^\mu \rangle_{\text{ren}}$ arises entirely from the subtraction terms, $S(x, x', s)_{\text{Ad}(4)}$. This means that $\langle \nabla_\mu j_A^\mu \rangle_{\text{ren}}$ is *independent of the choice of vacuum*. On the other hand, it is not difficult to see that only the terms with $k = 2$ in Equation (A5) produce a non-vanishing result. Additionally, terms involving derivatives of $E_2(x, x')$ must be disregarded because they involve five derivatives of the metric and hence are of the fifth adiabatic order. With all these considerations, Expression (A4) leads then to Formula (35).

References

1. Peskin, M.E.; Schroeder, D.V. *An Introduction to Quantum Field Theory*; Addison-Wesley: Reading, MA, USA, 1995.
2. Parker, L. The Creation of Particles in an Expanding Universe. Ph.D. Thesis, Harvard University, Cambridge, MA, USA, 1966.
3. Parker, L. Particle creation in expanding universes. *Phys. Rev. Lett.* **1968**, *21*, 562. [CrossRef]
4. Parker, L. Quantized fields and particle creation in expansing universes. I. *Phys. Rev.* **1969**, *183*, 1057–1068. [CrossRef]
5. Parker, L. Quantized fields and particle creation in expansing universes. II. *Phys. Rev. D* **1971**, *3*, 346–356. [CrossRef]
6. Adler, S.L. Axial vector vertex in spinor electrodynamics. *Phys. Rev.* **1969**, *177*, 2426. [CrossRef]
7. Bell, J.S.; Jackiw, R. A PCAC puzzle: $\pi^0 \to \gamma\gamma$ in the σ model. *Nuovo Cimento A* **1969**, *51*, 47. [CrossRef]
8. Schwinger, J. On gauge invariance and vacuum polarization. *Phys. Rev.* **1951**, *82*, 664. [CrossRef]
9. Dunne, G. Heisenberg-Euler effective Lagrangians: Basis and extensions. In *I. Kogan Memorial Volume, From Fields to Strings: Circumnavigating Theoretical Physics*; Shifman, M., Vainshtein, A., Wheater, J., Eds.; World Scientific: Singapore, 2007.
10. Ferreiro, A.; Navarro-Salas, J. Pair creation in electric fields, anomalies, and renormalization of the electric current. *Phys. Rev. D* **2018**, *97*, 125012. [CrossRef]
11. Kimura, T. Divergence of axial-vector current in the gravitational field. *Progress Theor. Phys.* **1969**, *42*, 1191. [CrossRef]
12. Abbott, B.P.; Jawahar, S.; Lockerbie, N.A.; Tokmakov, K.V. (LIGO Scientific Collaboration and Virgo Collaboration). Observation of Gravitational Waves from a Binary Black Hole Merger. *Phys. Rev. Lett.* **2016**, *116*, 061102. [CrossRef] [PubMed]

13. Abbott, B.P.; Jawahar, S.; Lockerbie, N.A.; Tokmakov, K.V. (LIGO Scientific Collaboration and Virgo Collaboration). GW170817: Observation of Gravitational Waves from a Binary Neutron Star Inspiral. *Phys. Rev. Lett.* **2017**, *119*, 161101. [CrossRef] [PubMed]
14. Jackson, J.D. *Classical Electrodynamics*; Wiley: New York, NY, USA, 1998.
15. Calkin, M.G. An invariance property of the free electromagnetic field. *Am. J. Phys.* **1965**, *33*, 958–960. [CrossRef]
16. Deser, S.; Teitelboim, C. Duality transformations of Abelian and non-Abelian gauge fields. *Phys. Rev.* **1976**, *D13*, 1592. [CrossRef]
17. Agullo, I.; Del Rio, A.; Navarro-Salas, J. Gravity and handedness of photons. *Int. J. Mod. Phys. D* **2017**, *26*, 1742001. [CrossRef]
18. Agullo, I.; Del Rio, A.; Navarro-Salas, J. Electromagnetic duality anomaly in curved spacetimes. *Phys. Rev. Lett.* **2017**, *118*, 111301. [CrossRef] [PubMed]
19. Agullo, I.; Del Rio, A.; Navarro-Salas, J. Classical and quantum aspects of electric-magnetic duality rotations in curved spacetimes. *Phys. Rev. D.* **2018**, *98*, 125001. [CrossRef]
20. Weinberg, S. *The Quantum Theory of Fields*; Cambridge University Press: Cambridge, UK, 1995; Volume 1.
21. Alvarez-Gaume, L.; Vazquez-Mozo, M. *An Invitation to Quantum Field Theory*; Springer: Heidelberg, Germany, 2012.
22. Srednicki, M. *Quantum Field Theory*; Cambridge University Press: Cambridge, UK, 2007.
23. Parker, L.; Toms, D. *Quantum Field Theory in Curved Spacetime*; Cambridge University Press: Cambridge, UK, 2009.
24. Agullo, I.; Parker, L. Non-gaussianities and the Stimulated creation of quanta in the inflationary universe. *Phys. Rev. D* **2011**, *83*, 063526. [CrossRef]
25. Agullo, I.; Parker, L. Stimulated creation of quanta during inflation and the observable universe. *Gen. Rel. Grav.* **2011**, *43*, 2541–2545. [CrossRef]
26. Gooth, J.; Niemann, A.C.; Meng, T.; Grushin, A.G.; Landsteiner, K.; Gotsmann, B.; Menges, F.; Schmidt, M.; Shekhar, C.; Süß, V.; et al. Experimental signatures of the mixed axial-gravitational anomaly in the Weyl semimetal NbP. *Nature* **2017**, *547*, 324. [CrossRef] [PubMed]
27. Leonhardt, U.; Philbin, T.G. General relativity in electrical engineering. *New J. Phys.* **2006**, *8*, 247. [CrossRef]
28. Pendry, J.B.; Schurig, D.; Smith, D.R. Controlling Electromagnetic Electromagnetic Fields. *Science* **2006**, *312*, 1780–1782. [CrossRef]

 © 2018 by the authors. Licensee MDPI, Basel, Switzerland. This article is an open access article distributed under the terms and conditions of the Creative Commons Attribution (CC BY) license (http://creativecommons.org/licenses/by/4.0/).

Article

On the Evolutionary Form of the Constraints in Electrodynamics

István Rácz [1,2]

[1] Faculty of Physics, University of Warsaw, Ludwika Pasteura 5, 02-093 Warsaw, Poland; istvan.racz@fuw.edu.pl or racz.istvan@wigner.mta.hu
[2] Wigner RCP, H-1121 Budapest, Konkoly Thege Miklós út 29-33, Hungary

Received: 12 November 2018; Accepted: 18 December 2018; Published: 22 December 2018

Abstract: The constraint equations in Maxwell theory are investigated. In analogy with some recent results on the constraints of general relativity, it is shown, regardless of the signature and dimension of the ambient space, that the "divergence of a vector field"-type constraint can always be put into linear first order hyperbolic form for which the global existence and uniqueness of solutions to an initial-boundary value problem are guaranteed.

Keywords: Maxwell theory; constraint equations; evolutionary equations

1. Introduction

The Maxwell equations, as we have known them since the seminal addition of Ampere's law by Maxwell in 1865, are [1]:

$$\nabla \times \mathbf{H} = \mathbf{J} + \partial_t \mathbf{D} \qquad \nabla \times \mathbf{E} + \partial_t \mathbf{B} = 0 \tag{1}$$

$$\nabla \cdot \mathbf{D} = q \qquad \nabla \cdot \mathbf{B} = 0, \tag{2}$$

where $\mathbf{E}$ and $\mathbf{B}$ are the macroscopic electric and magnetic field variables, which in a vacuum are related to $\mathbf{D}$ and $\mathbf{H}$ by the relations $\mathbf{D} = \epsilon_0 \mathbf{E}$ and $\mathbf{H} = \mu_0^{-1} \mathbf{B}$, where ϵ_0 and μ_0 are the dielectric constant and magnetic permeability and where q and $\mathbf{J}$ stand for charge and current densities, respectively.

The top two equations in Equation (1) express that the time-dependent magnetic field induces an electric field and also that the changing electric field induces a magnetic field even if there are no electric currents. Obviously, there have been plenty of brilliant theoretical, experimental and technological developments based on the use of these equations. Nevertheless, from time to time, some new developments (for a recent examples, see, for instance, [2,3]) have stimulated reconsideration of claims that previously were treated as text-book material in Maxwell theory.

In this short note, the pair of simple constraint equations on the bottom line in Equation (2) are the center of interest. These relations for the divergence of a vector field are customarily treated as elliptic equations. The main purpose of this letter is to show that by choosing basic variables in a geometrically-preferred way, the constraints in Equation (2) can also be solved as evolutionary equations. This also happens in the more complicated case of the constraints in general relativity [4].

Once the Maxwell Equations (1) and (2) are given, it is needless to explain in detail what is meant to be the ambient spacetime (tacitly, it is assumed to be the Minkowski spacetime) or the initial data surface (usually chosen to be a "$t = const$" hypersurface in Minkowski spacetime). As seen below, the entire argument, outlined in more detail in the succeeding sections, is very simple. In addition, it applies with almost no cost to a generic ambient space (M, g_{ab}), with a generic three-dimensional initial data surface Σ. We shall treat the generic case, i.e., solve the "divergence of a vector field type constraint",

$$\nabla \cdot \mathbf{L} = \ell, \quad \text{(in index notation)} \quad D_i L^i = \ell, \tag{3}$$

for a vector field **L** or (in index notation) L^i with a generic source ℓ, on a fixed, but otherwise arbitrary initial data surface, Σ. As an initial data surface can always be viewed as a time slice in an ambient spacetime, (M, g_{ab}), it is also straightforward to assign a Riemannian metric h_{ij} to Σ, the one induced by g_{ab} on Σ. In Equation (3) above, D_i stands for the unique torsion-free covariant derivative operator that is compatible with metric h_{ij}.

Note that for the Maxwell system, given by Equations (1) and (2), the two divergences of a vector field constraint decouple, so it suffices to solve them independently. Note also that it is easy to see that all the arguments presented in the succeeding subsections generalize to an arbitrary $n \geq 3$ dimension of Σ. Nevertheless, for the sake of simplicity, our consideration here will be restricted to the case of three-dimensional initial data surfaces.

Since the constraints are almost exclusively referred to as elliptic equations in text-books, one may question the point of putting them into evolutionary form. We believe that the appearance of time evolution in a Riemannian space could itself be of interest in its own right. Nevertheless, it is important to emphasize that there are valuable applications of the proposed new method. For instance, it may offer solutions to problems that are hard to solve properly in the standard elliptic approach. An immediate example of this sort arises in the initialization of the time evolution of point charges governed by the coupled Maxwell–Lorentz equations. As pointed out recently in [2], unless suitable additional conditions are applied in addition to the Maxwell constraints, the electromagnetic field develops singularities along the light cones emanating from the original positions of the point charges. It is important to be mentioned here that analogous problems arise in the context of the initialization of the time evolution of binary black hole configurations. In both cases, singularities are involved, which in the case of the Maxwell–Lorentz system are located at the point charges, whereas in the binary black hole case, at the spacetime singularities. The main task is to construct physically-adequate initial data specifications such that they are regular everywhere apart from these singularities. In the case of binary black hole configurations, this can be done by using the superposed Kerr–Schild metric, as an auxiliary ingredient in determining the freely-specifiable fields. Then, suitable "initial data" are chosen, in the distant radiation-dominated region, to the evolutionary form of the Hamiltonian and momentum constraints of general relativity. The desired initial data are completed finally by solving the corresponding initial value problem [5]. A completely analogous procedure is proposed to be used in initializing the time evolution of a pair of interacting point charges in Maxwell theory (a detailed outline of this proposal is given in Section 4). In this case, the "superposed" Liénard–Wiechert vector potentials are used, as an auxiliary ingredient to prescribe the freely-specifiable fields. In addition, suitable initial data have to be chosen with respect to the evolutionary forms of the constraints (see Equation (16) below) in a distant radiation-dominated region. The desired initial data can then be completed by solving the evolutionary form of the constraints Equation (2) as an initial value problem. It is remarkable that while in the conventional elliptic approach, some assumptions (in most cases tacit ones) are always used concerning the blow up rate (while approaching the singularities) of the constrained fields, no such fictitious "inner boundary condition" is applied anywhere in the proposed new method. It is indeed the evolutionary form of constraints itself that tells the constrained variables how they should evolve from their weak field values towards and up to the singularities.

An additional, and not the least important, potential advantage of the proposed new method is that it offers an unprecedented flexibility in solving the constraint equations. This originates from the fact that neither the choice of the underlying foliations of the three-dimensional initial data surface Σ, nor the choice of the evolutionary flow have any limitations. This makes the proposed method applicable to a high variety of problems that might benefit from this new approach to solving the constraints.

Another advantage of this new approach to the constraints is that, regardless of the choice of foliation and flow, the geometrically-preferred set of variables constructed in carrying out the main steps of the procedure always satisfy a linear first order symmetric hyperbolic equation. Considering the robustness of the approach, it is remarkable that, starting with the "divergence of a vector field constraint", the global existence of a unique smooth solution for the geometrically-preferred dependent variables (under suitable regularity conditions on the coefficients and source terms) is guaranteed for the linear first order symmetric hyperbolic equation (see, e.g., Subsection VIII.12.1 in [6]).

2. Preliminaries

The construction starts by choosing a three-dimensional initial data surface Σ with an induced Riemannian metric h_{ij} and its associated torsion-free covariant derivative operator D_i. Σ may be assumed to lie in an ambient space (M, g_{ab}) whose metric could have either a Lorentzian or Euclidean signature. More importantly, Σ will be assumed to be a topological product:

$$\Sigma \approx \mathbb{R} \times \mathscr{S}, \qquad (4)$$

where $\mathscr{S}$ could be of a two-surface with arbitrary topology. In the simplest practical case, however, $\mathscr{S}$ would have either a planar, cylindrical, toroidal, or spherical topology. In these cases, we may assume that there exists a smooth real function $\rho : \Sigma \to \mathbb{R}$ whose $\rho = const$ level sets give the $\mathscr{S}_\rho$ leaves of the foliation and that its gradient $\partial_i \rho$ does not vanish, apart from some isolated locations where the foliation may degenerate. (If, for instance, Σ has the topology $\mathbb{R}^3$, $\mathbb{S}^3$, $\mathbb{S}^2 \times \mathbb{R}$, or $\mathbb{S}^2 \times \mathbb{S}^1$ and it is foliated by topological two-spheres, then there exists one, two, or in the later two cases, no points of degeneracy at all. If point charges are involved, it may be preferable to place the associated physical singularities at the location of these degeneracies. Note also that we often write partial derivatives $\partial/\partial x^i$ in shorthand by ∂_i.)

The above condition guarantees (as indicated in Figure 1) that locally, Σ is smoothly foliated by a one-parameter family of $\rho = const$ level two-surfaces $\mathscr{S}_\rho$.

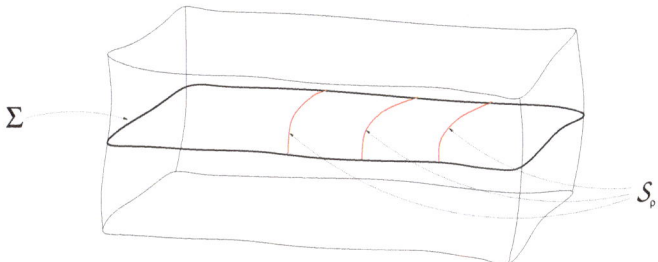

Figure 1. The initial data surface Σ foliated by a one-parameter family of two-surfaces $\mathscr{S}_\rho$ is indicated.

Given these leaves, the non-vanishing gradient $\partial_i \rho$ can be normalized to a unit normal $\widehat{n}_i = \partial_i \rho / \sqrt{h^{ij}(\partial_i \rho)(\partial_j \rho)}$, using the Riemannian metric h_{ij}. Raising the index according to $\widehat{n}^i = h^{ij} \widehat{n}_j$ gives the unit vector field normal to $\mathscr{S}_\rho$. The operator $\widehat{\gamma}^i{}_j$ formed from the combination of $\widehat{n}_i$ and $\widehat{n}^i$ and the identity operator $\delta^i{}_j$,

$$\widehat{\gamma}^i{}_j = \delta^i{}_j - \widehat{n}^i \widehat{n}_j \qquad (5)$$

projects fields on Σ to the tangent space of the $\mathscr{S}_\rho$ leaves.

We also apply flows interrelating the fields defined on the successive $\mathscr{S}_\rho$ leaves. A vector field ρ^i on Σ is called a flow if its integral curves intersect each of the leaves precisely once and it is normalized such that $\rho^i \partial_i \rho = 1$ holds everywhere on Σ. The contraction $\widehat{N} = \rho^j \widehat{n}_j$ of ρ^i with $\widehat{n}_i$ and its projection $\widehat{N}^i = \widehat{\gamma}^i{}_j \rho^j$ of ρ^i to the leaves are referred to as the "lapse" and "shift" of the flow, and we have:

$$\rho^i = \widehat{N} \widehat{n}^i + \widehat{N}^i. \qquad (6)$$

The inner geometry of the $\mathscr{S}_\rho$ leaves can be characterized by the metric:

$$\widehat{\gamma}_{ij} = \widehat{\gamma}^k{}_i \widehat{\gamma}^l{}_j h_{kl} \tag{7}$$

induced on the $\rho = const$ level surfaces. It is also known that a unique torsion-free covariant derivative operator $\widehat{D}_i$ associated with the metric $\widehat{\gamma}_{ij}$ acts on fields intrinsic to the $\mathscr{S}_\rho$ leaves, e.g., acting on the field $\mathbf{N}_l = \widehat{\gamma}^p{}_l N_p$ obtained by the projection of N_p according to:

$$\widehat{D}_i \mathbf{N}_j = \widehat{\gamma}^k{}_i \widehat{\gamma}^l{}_j D_k \left[\widehat{\gamma}^p{}_l N_p \right]. \tag{8}$$

It is straightforward to check that $\widehat{D}_i$ is indeed metric compatible in the sense that $\widehat{D}_k \widehat{\gamma}_{ij}$ vanishes.

Note also that the exterior geometry of the $\mathscr{S}_\rho$ leaves can be characterized by the extrinsic curvature tensor $\widehat{K}_{ij}$ and the acceleration $\dot{\hat{n}}_i$ of the unit normal, given by:

$$\widehat{K}_{ij} = \frac{1}{2} \mathscr{L}_{\hat{n}} \widehat{\gamma}_{ij} \quad \text{and} \quad \dot{\hat{n}}_i = \hat{n}^l D_l \hat{n}_i = -\widehat{D}_i \ln \widehat{N}, \tag{9}$$

where $\mathscr{L}_{\hat{n}}$ is the Lie derivative operator with respect to the vector field $\hat{n}^i$ and $\widehat{N}$ is the lapse of the flow.

3. The Evolutionary Form of the Constraints

This section is to put the divergence-type constraint Equation (3) into evolutionary form. This is achieved by applying a $2+1$ decomposition where, as we see below, the main conclusion is completely insensitive to the choice of the foliation and of the flow.

Consider first an arbitrary co-vector field L_i on Σ. By making use of the projector $\widehat{\gamma}^i{}_j$ defined in the previous section, we obtain:

$$L_i = \delta^j{}_i L_j = (\widehat{\gamma}^j{}_i + \hat{n}^j \hat{n}_i) L_j = \lambda \hat{n}_i + \mathbf{L}_i, \tag{10}$$

where the boldfaced variables λ and $\mathbf{L}_i$ are fields intrinsic to the individual $\mathscr{S}_\rho$ leaves of the foliation of Σ. They are defined via the contractions:

$$\lambda = \hat{n}^l L_l \quad \text{and} \quad \mathbf{L}_i = \widehat{\gamma}^j{}_i L_j. \tag{11}$$

By applying an analogous decomposition of $D_i L_j$, we obtain:

$$D_i L_j = \delta^k{}_i \delta^l{}_j D_k \left[\delta^p{}_l L_p \right] = (\widehat{\gamma}^k{}_i + \hat{n}^k \hat{n}_i)(\widehat{\gamma}^l{}_j + \hat{n}^l \hat{n}_j) D_k \left[(\widehat{\gamma}^p{}_l + \hat{n}^p \hat{n}_l) L_p \right], \tag{12}$$

which, in terms of the induced metric Equation (7), the associated covariant derivative operator, the extrinsic curvature, and the acceleration Equation (9), can be written as:

$$D_i L_j = \left[\widehat{D}_i \lambda + \hat{n}_i \mathscr{L}_{\hat{n}} \lambda \right] \hat{n}_j + \lambda \left(\widehat{K}_{ij} + \hat{n}_i \dot{\hat{n}}_j \right) + \widehat{D}_i \mathbf{L}_j - \hat{n}_i \hat{n}_j \left(\dot{\hat{n}}^l \mathbf{L}_l \right)$$
$$+ \left\{ \hat{n}_i \mathscr{L}_{\hat{n}} \mathbf{L}_j - \hat{n}_i \mathbf{L}_l \widehat{K}^l{}_j - \hat{n}_j \mathbf{L}_l \widehat{K}^l{}_i \right\}. \tag{13}$$

By contracting the last equation with the inverse $h^{ij} = \widehat{\gamma}^{ij} + \hat{n}^i \hat{n}^j$ of the three-metric h_{ij} on Σ, we obtain:

$$D^l L_l = h^{ij} D_i L_j = (\widehat{\gamma}^{ij} + \hat{n}^i \hat{n}^j) D_i L_j = \mathscr{L}_{\hat{n}} \lambda + \lambda \left(\widehat{K}^l{}_l \right) + \widehat{D}^l \mathbf{L}_l + \dot{\hat{n}}^l \mathbf{L}_l. \tag{14}$$

By virtue of Equation (3) and in accord with the last equation, it is straightforward to see that the divergence of a vector field constraint can be put into the form:

$$\mathscr{L}_{\hat{n}} \lambda + \lambda \left(\widehat{K}^l{}_l \right) + \widehat{D}^l \mathbf{L}_l + \dot{\hat{n}}^l \mathbf{L}_l = \ell. \tag{15}$$

Now, by choosing arbitrary coordinates (x^2, x^3) on the $\rho = const$ leaves and by Lie dragging them along the chosen flow ρ^i, coordinates (ρ, x^2, x^3) adapted to both the foliation $\mathscr{S}_\rho$ and the flow $\rho^i = (\partial_\rho)^i$ can be introduced on Σ. In these coordinates, Equation (15) takes the strikingly simple form in terms of the lapse and shift of the flow,

$$\partial_\rho \lambda - \widehat{N}^K \partial_K \lambda + \lambda \, \widehat{N} \, (\widehat{K}^L{}_L) + \widehat{N} [\widehat{D}_L \mathbf{L}^L + \hat{n}_L \mathbf{L}^L] = \ell. \tag{16}$$

Some remarks are now in order. First, Equation (16) is a scalar equation whereby it is natural to view it as an equation for the scalar part $\lambda = \hat{n}_i L^i$ of the vector field L^i on Σ and to solve it for λ. All the coefficients and source terms in Equation (16) are determined explicitly by freely specifying the fields $\mathbf{L}^L$ and ℓ, whereas the metric h_{ij} and its decomposition in terms of the variables $\widehat{N}, \widehat{N}^I, \widehat{\gamma}_{IJ}$, is also known throughout Σ. Thus, Equation (16) can be solved for λ. Note that Equation (16) is manifestly independent of the choice made for the foliation and flow and also that Equation (16) is always a linear hyperbolic equation for λ, with ρ "playing the role of time".

4. A Simple Example

Though the results in the previous section are mathematically all robust, it would be pointless to have the proposed evolutionary form of the constraints unless one could apply it in solving certain problems of physical interest. In order to get some hints of how the proposed techniques work, this section is to give an outline of a construction that could be used to get meaningful initialization of the time evolution of a pair of moving point charges in Maxwell theory.

Recall first that accelerated charges are known to emit electromagnetic radiation. An interesting particular case is when the radiation is emitted by a pair of point charges moving as dictated by their mutual electromagnetic field. To start off, choose the $t = 0$ time slice in a background Minkowski spacetime. This time slice itself is a three-dimensional Euclidean space $\mathbb{R}^3$ that can be endowed with the conventional Cartesian coordinates (x, y, z) as a three-parameter family of inertial observers has already been chosen in the ambient Minkowski background. Assume that on this time slice, the two point charges are located on the $y = z = 0$ line at $x = \pm a$ (with some $a > 0$), each moving with some initial speed. Choose then a one-parameter family of confocal rotational symmetric ellipsoids:

$$\begin{aligned} x &= a \cdot \cosh \rho \cdot \cos \chi \\ y &= a \cdot \sinh \rho \cdot \sin \chi \cdot \cos \varphi \\ z &= a \cdot \sinh \rho \cdot \sin \chi \cdot \sin \varphi \,. \end{aligned} \tag{17}$$

It is straightforward to check that $\mathbb{R}^3$ gets to be foliated by the $\rho = const$ level surfaces, which are confocal rotational ellipsoids:

$$\frac{x^2}{a^2 \cdot \cosh^2 \rho} + \frac{y^2 + z^2}{a^2 \cdot \sinh^2 \rho} = 1, \tag{18}$$

with focal points $f_+ = (a, 0, 0)$ and $f_- = (-a, 0, 0)$. Note also that each member of the two-parameter family of curves determined by the relations $\chi = const, \varphi = const$, with $0 < \chi \leq 2\pi$, parameterized by $\rho \, (\geq 0)$, intersect $\rho = const$ level surfaces precisely once. The introduced new coordinates (ρ, χ, φ) cover the complement of the two focal points in $\mathbb{R}^3$. Choose this complement as our initial data surface Σ. These coordinates, adopted with respect to the $\rho : \Sigma \to \mathbb{R}$ foliation and to the flow vector field ρ^i on Σ, are such that ρ^i is parallel to the $\chi = const, \varphi = const$ coordinate lines and is normalized such that $\rho^i(\partial_i \rho) = 1$. The pertinent laps and shift, $\widehat{N}$ and $\widehat{N}^i$, of this coordinate bases vector $\rho^i = (\partial_\rho)^i$ can also be determined as described in Section 2.

Following then a strategy analogous to the one applied in getting the binary black hole initial data in general relativity [5], one may proceed as follows. By superposing the Liénard–Wiechert vector potentials relevant for the individual point charges, moving with certain initial speeds, determine first the corresponding auxiliary Faraday tensor $^{(aux)}F_{ab}$. Restrict it to the $t = 0$ initial data surface, and extract there the auxiliary electric $^{(aux)}\mathbf{E}$ and magnetic $^{(aux)}\mathbf{B}$ fields. These electric and magnetic parts of $^{(aux)}F_{ab}$ are meant to be defined with respect to the aforementioned three-parameter family of static observers (moving in the background Minkowski spacetime with four velocity $u^a = (\partial_t)^a$). Split these vector fields, as described at the beginning of Section 3, into scalar and two-dimensional vector parts; we get $^{(aux)}E_i = {^{(aux)}\varepsilon}\,\hat{n}_i + {^{(aux)}\mathcal{E}_i}$ and $^{(aux)}B_i = {^{(aux)}\beta}\,\hat{n}_i + {^{(aux)}\mathcal{B}_i}$, respectively. The two-dimensional vector parts $^{(aux)}\mathcal{E}_A$ and $^{(aux)}\mathcal{B}_A$, of the auxiliary electric $^{(aux)}\mathbf{E}$ and magnetic $^{(aux)}\mathbf{B}$ fields, are well-defined smooth fields on Σ. As they encode important information about the momentary kinematical content of the considered system, e.g., the initial speeds and locations of the involved point charges, the fields $^{(aux)}\mathcal{E}_A$ and $^{(aux)}\mathcal{B}_A$ are used as the freely-specified part of data throughout Σ. Once this has been done, the radiation content of the initial data, for the physical $\mathbf{E}$ and $\mathbf{B}$, in the far zone has to be introduced by choosing—based on measurements, expectations, and/or intuition—two smooth functions, $_{(0)}\varepsilon$ and $_{(0)}\beta$, on a level surface $\rho = \rho_0$ (for some sufficiently large real value of ρ_0) in Σ. These are the initial data with respect to the pertinent forms of Equation (16) that can be deduced—as described in Sections 2 and 3—from the constraints equations in Equation (2).

Remarkably, for arbitrarily small values of $\epsilon > 0$, unique smooth solutions ε and β to the (decoupled) evolutionary form of the constraint equations exist in the region bounded by the $\rho = \rho_0$ and $\rho = \epsilon$ level surfaces (one could integrate the equations also outwards, with respect to $\rho = \rho_0$; nevertheless, if one is interested in the behavior of the initial data in the near zone region, then the aforementioned domain is the relevant one). The corresponding unique smooth solutions smoothly extend onto Σ, even in the $\epsilon \to 0$ limit, in spite of the fact the solutions are known to blow up at the focal points where the point charges are located initially. Using the unique smooth solution ε and β, corresponding to the choices made for the initial data $_{(0)}\varepsilon$ and $_{(0)}\beta$ at $\rho = \rho_0$, the initialization of the physical electric and magnetic fields is given as $E_i = \varepsilon\,\hat{n}_i + {^{(aux)}\mathcal{E}_i}$ and $B_i = \beta\,\hat{n}_i + {^{(aux)}\mathcal{B}_i}$, respectively. Note that they will differ from $^{(aux)}E_i$ and $^{(aux)}B_i$ as the initial data $_{(0)}\varepsilon$ and $_{(0)}\beta$, for the pertinent forms of Equation (16), were chosen to differ from $^{(aux)}\varepsilon$ and $^{(aux)}\beta$. More importantly, once the electric and magnetic fields $\mathbf{E}$ and $\mathbf{B}$ are initialized in the way prescribed above, the conventional time evolution equations of the coupled Maxwell–Lorentz system (including the two ones in Equation (1)) relevant for the pair of interacting point charges should be solved (possible by numerical means). Notably, due to the above outlined initialization, the radiation that will emerge from the consecutive accelerating motion of the pair of point charges is guaranteed to be consistent with the radiation imposed, by specifying initial data $_{(0)}\varepsilon$ and $_{(0)}\beta$, at the $\rho = \rho_0$ level surface located in the far zone.

5. Final Remarks

By virtue of the main result of this note, the "divergence of a vector-type constraint" can always be solved as a linear first order hyperbolic equation for the scalar part of the vector variable under consideration. As was emphasized in the Introduction, robust mathematical results guarantee the global existence of unique smooth solutions (under suitable regularity conditions on the coefficients and source terms) to the linear first order symmetric hyperbolic equations of the form of Equation (16).

The real strength of the proposed method emanates from the freedom we have in choosing the applied $1+2$ decomposition. As we saw no matter how the foliation, determined by a smooth real function $\rho : \Sigma \to \mathbb{R}$, and the flow vector field ρ^i are chosen, the pertinent ρ coordinate will always play the role of time in the pertinent evolutionary form of the constraints. In order to provide some evidence concerning the capabilities and some of the prosperous features of the proposed method, the basic steps of initializing the time evolution of a pair of interacting point charges were also outlined. This simple example should also provide a clear manifestation of the agreement, which always comes along with the use of the proposed evolutionary form of the constraints in electrodynamics.

Funding: This project was supported by the POLONEZ program of the National Science Centre of Poland, which has received funding from the European Union's Horizon 2020 research and the innovation program under the Marie Skłodowska-Curie Grant Agreement No. 665778.

Acknowledgments: The author is deeply indebted to Jeff Winicour for his careful reading and for a number of helpful comments and suggestions.

Conflicts of Interest: The author declares no conflict of interest.

References

1. Jackson, J.D. *Classical Electrodynamics*, 3rd ed.; John Wiley & Sons, Inc.: Hoboken, NJ, USA, 1999.
2. Deckert, D.-A.; Hartenstein, V. On the initial value formulation of classical electrodynamics. *J. Phys. A Math. Theor.* **2016**, *49*, 445202. [CrossRef]
3. Medina, R.; Stephany, J. Momentum exchange between an electromagnetic wave and a dispersive medium. *arXiv* **2018**, arXiv:1801.09323.
4. Rácz, I. Constrains as evolutionary systems. *Class. Quant. Grav.* **2016**, *33*, 015014. [CrossRef]
5. Rácz, I. A simple method of constructing binary black hole initial data. *Astron. Rep.* **2018**, *62*, 953–958.
6. Choquet-Bruhat, Y. *Introduction to General Relativity, Black Holes and Cosmology*; Oxford University Press: Oxford, UK, 2015.

© 2018 by the author. Licensee MDPI, Basel, Switzerland. This article is an open access article distributed under the terms and conditions of the Creative Commons Attribution (CC BY) license (http://creativecommons.org/licenses/by/4.0/).

Article

Maxwell Electrodynamics in Terms of Physical Potentials

Parthasarathi Majumdar [1,*] and Anarya Ray [2]

1. Indian Association for the Cultivation of Science, Kolkata 700032, India
2. Department of Physics, Presidency University, Kolkata 700032, India
* Correspondence: bhpartha@gmail.com

Received: 11 June 2019; Accepted: 12 July 2019; Published: 14 July 2019

Abstract: A fully relativistically covariant and manifestly gauge-invariant formulation of classical Maxwell electrodynamics is presented, purely in terms of gauge-invariant potentials without entailing any gauge fixing. We show that the inhomogeneous equations satisfied by the physical scalar and vector potentials (originally discovered by Maxwell) have the same symmetry as the isometry of Minkowski spacetime, thereby reproducing Einstein's incipient approach leading to his discovery of special relativity as a spacetime symmetry. To arrive at this conclusion, we show how the Maxwell equations for the potentials follow from stationary electromagnetism by replacing the Laplacian operator with the d'Alembertian operator, while making all variables dependent on space and time. We also establish consistency of these equations by deriving them from the standard Maxwell equations for the field strengths, showing that there is a unique projection operator which projects onto the physical potentials. Properties of the physical potentials are elaborated through their iterative Nöther coupling to a charged scalar field leading to the Abelian Higgs model, and through a sketch of the Aharonov–Bohm effect, where dependence of the Aharonov–Bohm phase on the physical vector potential is highlighted.

1. Introduction

The standard textbook formulation of Maxwell electrodynamics, in vacua with sources, entails linear first order partial differential equations for electric and magnetic field strengths $\vec{E}$ and $\vec{B}$. Conventionally, the equations for these field strengths are first cast in terms of the scalar and vector *potentials*, ϕ and $\vec{A}$. The resulting second order equations for the potentials are found to be noninvertible because of the *gauge ambiguity* of the potentials—addition of gradients of arbitrary (gauge) functions to any solution generates an equivalence class of solutions for the potentials, related by local gauge transformations. All the gauge potentials in a gauge-equivalent class give the same electromagnetic field strengths. This gauge ambiguity has often led people to consider gauge potentials as *unphysical*, in comparison to the 'physical' (gauge-invariant) field strengths. The standard procedure for getting to the solutions is to 'gauge fix' the potentials, i.e., impose subsidiary conditions on them so that the ambiguity may be resolved. There is a nondenumerably infinite set of such subsidiary 'gauge conditions', each one as ad hoc as the other, and none with any intrinsic physical relevance. This entire approach, tenuous as it is, avoids facing up to the central issue: Why are the equations for the potentials noninvertible in the first place? Is Nature so unkind as to provide us with unique gauge-invariant equations for quantities which themselves are infinitely ambiguous? The answer is an emphatic *No!*

In his succinctly beautiful history of the Maxwell equations of electrodynamics, Nobel Laureate theoretical physicist C. N. Yang [1] recalls how Faraday first identified the concept of the 'electrotonic state' as the origin of the induced electromotive force, purely as a result of his extraordinary experimental research and physical intuition. The idea of the vector potential was introduced by

Thomson (Lord Kelvin) in 1851, ostensibly as a solution of $\nabla \cdot \vec{B} = 0$. Five years later, in a brilliant identification of Thomson's vector potential with Faraday's electrotonic state, Maxwell wrote down, for the first time ever, the equation $\vec{E} = -\dot{\vec{A}}$, which led him to his Law VI: *The electromotive force on any element of a conductor is measured by the instantaneous rate of change of the electrotonic intensity on that element, whether in magnitude or direction.* Yang further writes: "The identification of Faraday's elusive idea of the electrotonic state (or electrotonic intensity, or electrotonic function) with Thomson's vector potential is, in my opinion, the first great conceptual breakthrough in Maxwell's scientific research...", also, "Indeed, the concept of the vector potential remained central in Maxwell's thinking throughout his life."

From our standpoint, it is inconceivable that such an outstanding experimentalist as Faraday would focus on a concept which we call the vector potential, if indeed it is 'unphysical', as often perceived nowadays among a certain group of physicists. Likewise, Maxwell's preoccupation with the same concept would have been a continuation of an illusory pursuit if the vector potential is indeed unphysical. Interestingly, Maxwell himself was apparently quite aware of the gauge ambiguity of the Maxwell equations for $\vec{E}$ and $\vec{B}$ as mentioned above, but according to Yang [1], on the issue of gauge-fixing Maxwell was silent: 'He did not touch on that question, but left it completely indeterminate.' This is where we speculate on the reason: Maxwell was perhaps aware that his equations themselves provided, in today's parlance, a *unique projection operator* which projects onto the physical part of the vector potential. Clearly, $\vec{E}$ and $\vec{B}$ depend *solely on this physical part of the vector potential* and are quite independent of the unphysical *pure gauge* part. The manner in which the projection operator isolates the gauge-invariant physical part of the vector potential, can, of course be reproduced by a gauge choice as well—however, such a choice is *by no means* essential. Gauge choices (or gauge-fixings) merely constrain the unphysical part of the gauge potential, leaving the gauge-invariant physical part quite untouched, as they must.

To reiterate, the reason that the equations for the potentials are noninvertible in the first place is because their intrinsic analytic structure involves a *projection* operator which has a nontrivial kernel of unphysical 'pure gauge' vector fields! This simple observation renders any 'gauge fixing' superfluous, since it is now obvious that the equations are to be interpreted in terms of projected *physical, gauge-invariant* potentials not belonging to the kernel of the projection operator, hence obeying very simple wave equations that are immediately uniquely invertible without the need for any imposition of additional 'gauge conditions'. We find it surprising that this simple fact has not been clarified in any of the number of currently popular textbooks on classical electrodynamics. From a physical standpoint, this approach, in contrast to the standard one based on the field strengths, immediately divulges the essence of electromagnetism as the theory of electromagnetic *waves* under various circumstances. All other field configurations (in vacuo) can be easily explained once the propagation and generation of electromagnetic waves is understood in terms of the physical potentials.

There is another lacuna in extant textbook treatments of Maxwell electrodynamics—the absence of a fully relativistically covariant formulation of the subject ab initio. Special relativity is intrinsically embedded in Maxwell electrodynamics with charge and current sources in empty space, as was discovered by Einstein in 1905 [2]. If, as per standard practice, the fundamental equations are written in terms of the electric and magnetic fields, the relativistic invariance of these equations is far from obvious. This emerges only after some effort is given to relate the electric and magnetic fields in different inertial reference frames connected by Lorentz boosts. In contrast, if the equations are cast in terms of the *physical* electromagnetic scalar and vector potentials introduced by Maxwell, then these potentials and the equations they obey can be easily combined to yield a structure that is *manifestly* invariant under Lorentz boosts as well as spatial rotations, i.e., the full Lorentz transformations. Given that it is easier always to compute four, rather than six, field components for given source charge and current densities, it stands to reason to begin any formulation of electrodynamics from the (physical) potentials, rather than the field strengths.

Despite its antiquity, a formulation of classical electrodynamics that, from the outset, is fully relativistically covariant, is somehow not preferred in the very large number of excellent textbooks currently popular, with perhaps the sole exception of [3]. Even so, the issue of the gauge ambiguity and the full use of electromagnetic potentials rather than fields has not been dealt with adequately, even in this classic textbook. Thus, while relativistically covariant Lienard–Wiechert potentials describing the solution of Maxwell's equations due to an arbitrarily moving relativistic point charge have been obtained, the corresponding field strengths and the radiative energy–momentum tensor have not been given such a manifestly covariant treatment. The more widely used textbook [4] also fills this gap only in part. Since special relativity is so intrinsic to Maxwell vacuum electrodynamics with sources, it is only befitting that the entire formalism exhibit this symmetry explicitly. The subtle interplay with gauge invariance is also a hallmark of this theory, which forms the basis of our current understanding of the fundamental interactions of physics.

We end this introduction with the disclaimer—this paper is exclusively on Maxwell electrodynamics. As such, it does not discuss theoretically interesting generalizations involving magnetic charges and large gauge transformations. Interesting as these ideas are, there is no observational evidence yet that they are applicable to the physical universe, so these ideas remain within the domain of speculation. Of course, the moment a magnetic monopole is observed as an asymptotic state, our paper stands to be immediately falsified. This, however, is not a lacuna of the paper, rather it is its strength that it 'sticks its neck out' so to speak, in contrast with the plethora of theoretical papers whose veracity or relevance vis-a-vis the physical universe remains forever in doubt. Regarding the generalization of electrodynamics with both electric and magnetic charges, the construction of a local field theory is still not without issues. Whether this is a hint from Nature about the relevance of such ideas, still remains unclear. In our favor, the entire description being in terms of a single 4-vector potential has a virtue: The electric and magnetic aspects are actually *unified* in this description. If magnetic monopoles were present, this unification would actually be absent, in favor of a duality symmetry—the electric–magnetic duality. It is not unlikely that even though this duality symmetry is dear to some theoretical physicists, Nature does not make use of it, if the evidence so far is to be taken into account. In regard to large gauge transformations, recent work on asymptotic symmetries of flat spacetime has led to interesting issues regarding electrodynamic gauge transformations, which may serve as research topics for the future.

This paper is structured as follows: In Section 2 we first exhibit gauge-invariant physical potentials for stationary electromagnetism (electrostatics and magnetostatics) and show how they satisfy *identical* equations underlining their inherent unity. We then generalize this to full electrodynamics with a neat substitution, and show that the standard Maxwell equations for field strengths *emerge* from these. In the next section, we demonstrate the invariance of the Maxwell equations for the physical potentials under Lorentz transformations, characterized by a 4×4 matrix Λ which includes both spatial rotations and Lorentz boosts. We argue that this symmetry of Maxwell electrodynamics is also the isometry of the Minkowski metric of flat spacetime. Next, we complete the circle by showing how the equations for the physical potentials can be derived covariantly from the covariant Maxwell equations for the field strengths and exhibit the form of the projection operator, which enables this projection onto physical potentials. Then, in Section 4, we show how the physical potentials can be coupled gauge-invariantly to charged scalar fields through the technique of *iterative Nöther coupling*, leading to the classical Abelian Higgs model. We also provide a sketch to show that the Aharonov–Bohm phase is a functional of the physical potentials alone, completely independent of the unphysical pure gauge part. We conclude in Section 5.

2. Gauge-Invariant Physical Potentials

2.1. Stationary Electromagnetism and Physical Potentials

Electrostatics is described by the equations

$$\begin{aligned} \nabla \cdot \vec{E} &= \rho, \\ \nabla \times \vec{E} &= 0, \end{aligned} \qquad (1)$$

where constants appropriate to choice of units have been set to unity. The solution of the second line of Equation (1) is $E = -\nabla \phi$, where ϕ is the scalar potential. ϕ is unique modulo an added constant. Substitution of this solution into the first line of Equation (1) results in the Poisson equation for ϕ:

$$\nabla^2 \phi_P = -\rho. \qquad (2)$$

Here, we have added a subscript P to ϕ to emphasize its physicality. Equation (2) has the (inhomogeneous) solution:

$$\phi_P(\vec{x}) = \int d^3x' \frac{\rho(\vec{x}')}{|\vec{x} - \vec{x}'|}. \qquad (3)$$

Likewise, stationary magnetism begins with the equation:

$$\begin{aligned} \nabla \cdot \vec{B} &= 0, \\ \nabla \times \vec{B} &= \vec{j}. \end{aligned} \qquad (4)$$

Apparently, the equations in (4) have nothing in common with their electrostatic counterpart (1), leading to the idea espoused in textbooks that electrostatics and magnetostatics are two separate disciplines. We now show that this idea is not correct at all, once the equations are transcribed in terms of the *physical* scalar and vector potentials.

Solving the first of the Equation (4) in terms of the vector potential $\vec{A}$: $\vec{B} = \nabla \times \vec{A}$, and substituting that in the second line of (4) yields

$$\nabla(\nabla \cdot \vec{A}) - \nabla^2 \vec{A} = \vec{j}. \qquad (5)$$

Now, Equation (5) cannot be solved uniquely for $\vec{A}$ because of the gauge ambiguity—there is an infinity of *gauge-equivalent* solutions for any $\vec{A}$ that satisfies (5). Usually, the way out of this ambiguity, as mentioned in almost all textbooks, is *gauge fixing*—choose $\nabla \cdot \vec{A}$ to be any specific function $f(\vec{x})$, leading to the vectorial Poisson equation which can be solved immediately with suitable boundary conditions, for every given source $\vec{j}$. The choice $f(\vec{x}) = 0$ is called the Coulomb gauge in some textbooks. But this is *not* what we propose to do here!

Instead, consider the Fourier transform of (5):

$$-\vec{k}[\vec{k} \cdot \vec{\tilde{A}}(\vec{k})] + k^2 \vec{\tilde{A}}(\vec{k}) = \vec{\tilde{j}}(\vec{k}), \qquad (6)$$

where $k^2 = k_a k_a$, $a = 1, 2, 3$. Switching to component notation, Equation (6) can be rewritten as

$$k^2 \mathcal{P}_{ab} \tilde{A}_b = \tilde{j}_b, \qquad (7)$$

where the *projection operator* $\mathcal{P}_{ab}$ is defined as

$$\mathcal{P}_{ab} \equiv \delta_{ab} - \frac{k_a k_b}{k^2}, \quad \mathcal{P}_{ab} \mathcal{P}_{bc} = \mathcal{P}_{ac}. \qquad (8)$$

Clearly, in defining this projection operator, we have chosen $k^2 \neq 0$, i.e., $\vec{k}$ is a nonzero vector. We shall comment on the zero vector situation below.

Define now the projected vector potential $\vec{A}_P$ through the equation

$$\tilde{A}_{Pa} \equiv \mathcal{P}_{ab} \tilde{A}_b . \tag{9}$$

This projected vector potential satisfies two very important properties: First of all, it is *gauge-invariant* and hence, *physical*! This follows from the fact that $k_a \mathcal{P}_{ab} = 0 \ \forall \ \vec{k} \neq 0$. Secondly, this last relation also implies that the projected vector potential satisfies

$$\vec{k} \cdot \vec{A}_P = 0 \Rightarrow \nabla \cdot \vec{A}_P = 0 \tag{10}$$

automatically, without having to make any gauge choice! Note that this is *not* the so-called Coulomb gauge choice. It is rather the consequence of defining the projected vector potential A_P using the projection operator that occurs already in Maxwell's equations for magnetostatics. No extraneous choice needs to be made for this physical projection—it is a unique projection. In fact, it now becomes clear as to why Ampere's law, written out in terms of the full vector potential, is not invertible: This equation involves the projection operator $\mathcal{P}_{ab}$, projection operators are not invertible because they have a nontrivial kernel—here, it is the set of pure gauge configurations expressible as ∇a for arbitrary scalar functions a. Once the projected vector potential A_P is defined, it satisfies the vector Poisson equation

$$\nabla^2 \vec{A}_P = -\vec{j}, \tag{11}$$

with the solution (in Fourier space)

$$\vec{A}_P = \frac{\vec{j}}{k^2} . \tag{12}$$

In position space, this solution is

$$\vec{A}_P(\vec{x}) = \int d^3 x' \frac{\vec{j}(\vec{x}')}{|\vec{x} - \vec{x}'|}, \tag{13}$$

which clearly shows that, unlike the magnetic field, the vector potential tracks the current producing it.

There is an issue of a residual gauge invariance for $k^2 = 0$ which has been excluded from our earlier discussion. If $k^2 = 0$, in position space, this would be taken to imply that the projected vector $\vec{A}_P$ has the residual ambiguity under the gauge transformation $\vec{A}_P \to \vec{A}_P + \nabla \omega$ with $\nabla^2 \omega = 0$. However, from the uniqueness of solutions of Laplace equation, we know that choosing the boundary condition $\omega = constant$ at spatial infinity implies that $\omega = const$ everywhere, thereby precluding any such residual ambiguity for the physical $\vec{A}_P$.

Summarizing stationary electromagnetism, we have, as the fundamental equations in terms of the physical potentials,

$$\nabla^2 \phi_P = -\rho. \tag{14}$$
$$\nabla^2 \vec{A}_P = -\vec{j}. \tag{15}$$
$$\nabla \cdot \vec{A}_P = 0 . \tag{16}$$

The great mathematical unity between the physical electric potential and magnetic vector potentials, in terms of the equations they satisfy, need hardly be overemphasized. It is also perhaps the most succinct manner of presentation of stationary electromagnetism.

2.2. Formulation of Full Electrodynamics in Terms of Physical Potentials

The passage from the stationary Equations (14)–(16) to the full time-dependent equations for the physical potentials follows the exceedingly simple rule: Allow all functions of space to now be functions of space and time, i.e., $\vec{x}$ and t—and make the simple change $\nabla^2 \to \Box^2$ in Equations (14) and (15), where, $\Box^2 \equiv \nabla^2 - (\partial^2/\partial t^2)$ is the d'Alembertian operator, and we are using units such that $c = 1$. Further, the divergence-free condition (16) is to be augmented by the 'spacetime' divergence-free condition $(\partial \phi_P / \partial t) + \nabla \cdot \vec{A} = 0$, leading to the Maxwell equations for the physical electromagnetic scalar and vector potentials:

$$\begin{aligned} \Box^2 \phi_P &= -\rho, \\ \Box^2 \vec{A}_P &= -\vec{j}, \\ \frac{\partial \phi_P}{\partial t} + \nabla \cdot \vec{A}_P &= 0, \end{aligned} \quad (17)$$

where all other constants are absorbed into redefinitions of ρ and $\vec{j}$.

Equation (17) can be motivated physically ab initio from the most important characteristic of electrodynamics, namely, that they must yield electromagnetic waves traveling through empty space. Indeed, the top two equations in (17) are nothing but *inhomogeneous* wave equations, recalling that the d'Alembertian is the *wave* operator. The last equation is a special characteristic property of the physical potentials ϕ_P, $\vec{A}_P$, which is relevant to ensure that the electromagnetic waves in empty space have transverse polarization. We think that electromagnetic waves constitute the most important physical property of electrodynamics, and our formulation of Maxwell's theory in terms of the potentials brings out this characteristic immediately and without the need for extraneous manipulations. Thus, one can start with the formulation in terms of Equation (17) by pointing out that they epitomize electromagnetic waves, and form the basis of what we see, of ourselves, of worlds outside ours and also in terms of constructing theories of fundamental interactions.

The standard Maxwell equations for the electric and magnetic field strengths, $\vec{E}$ and $\vec{B}$, result from (17) immediately upon using Maxwell's definition of Faraday's 'electrotonic' state, as defined in the introduction: $\vec{E} \equiv -\partial \vec{A}_P / \partial t - \nabla \phi_P$ and Thomson's definition $\vec{B} \equiv \nabla \times \vec{A}_P$. One obtains

$$\begin{aligned} \nabla \cdot \vec{E} &= \rho, \\ \nabla \cdot \vec{B} &= 0, \\ \nabla \times \vec{E} &= -\frac{\partial \vec{B}}{\partial t}, \\ \nabla \times \vec{B} &= \vec{j} + \frac{\partial \vec{E}}{\partial t}. \end{aligned} \quad (18)$$

3. Special Relativity as a Symmetry of Maxwell's Equations

3.1. Manifest Lorentz Symmetry

Observe that these equations can be combined into the 4-vector potential $\mathbf{A}_P$ with components $A_P^\mu, \mu = 0, 1, 2, 3$, with $A_P^0 = -A_{P0} = \phi$, and $A_P^m = A_{P,m}, m = 1, 2, 3$, where $\vec{A}_P = \{A_P^m | m = 1, 2, 3\}$. Similarly, the charge and current densities can be combined into a current density 4-vector $\mathbf{J}$ with components J^μ ($\mu = 0, 1, 2, 3) = (\{\rho, j^a\}, a = 1, 2, 3)$.

Writing $\partial_\mu \equiv \partial/\partial x^\mu$, Equation (17) can now be summarized as

$$\begin{aligned} \Box^2 A_P^\mu &= -J^\mu, \\ \partial_\mu A_P^\mu &= 0. \end{aligned} \quad (19)$$

Raising and lowering of indices are effected by the Minkowski spacetime invariant metric tensor $\eta_{\mu\nu} = \eta^{\mu\nu} = \mathrm{diag}(-1, 1, 1, 1)$.

It is to be noted that Equations (17) and (19) are *not* gauge-fixed versions of equations for potentials corresponding to standard Maxwell equations, even though they look enticingly similar. In other words, the second of the equations in (19) is *not a gauge choice*, but a compulsion from Nature. We shall make this clear shortly. Thus, these equations are to be treated at the same physical footing as the standard Maxwell equations for the field strengths, containing the same physical information as the latter, without any ambiguity.

Observe now that the d'Alembertian operator $\Box^2 \equiv \eta^{\mu\nu}\partial_\mu\partial_\nu$ is *invariant* under the transformation $\partial_\mu \to \partial'_\mu = \Lambda_\mu{}^\nu \partial_\nu$, provided the transformation matrix Λ satisfies

$$\eta_{\mu\nu} \Lambda^\mu{}_\rho \Lambda^\nu{}_\sigma = \eta_{\rho\sigma}. \tag{20}$$

It follows that both lines of Equation (19) are invariant under these transformations, if $A'^\mu(x') = \Lambda^\mu{}_\nu A^\nu(x)$, $J'^\mu(x') = \Lambda^\mu{}_\nu J^\nu(x)$, and $x'^\mu = \Lambda^\mu{}_\nu x^\nu$, with the transformation matrices $\Lambda^\mu{}_\nu$ satisfying (20). As expected, the coordinate transformations leave invariant the squared invariant interval in Minkowski spacetime $ds^2 = \eta_{\mu\nu} dx^\mu dx^\nu$. Thus, the transformations that leave the equations of electrodynamics invariant are precisely the same transformations that constitute a *symmetry* of Minkowski spacetime. It is obvious that these are the full Lorentz transformations, including spatial rotations and Lorentz boosts, e.g., if $\Lambda^0{}_0 = 1$, $\Lambda^0{}_m = 0$, the remaining 3×3 submatrix constitutes the *orthogonal* transformation matrix corresponding to rotations in 3-space. Likewise, if $\Lambda^0{}_0 = \gamma = \Lambda^1{}_1$; $\Lambda^0{}_1 = -\beta\gamma = \Lambda^1{}_0$ etc., that constitutes a Lorentz boost in the $+x^1$ direction. The Lorentz factor $\gamma = (1 - \beta^2)^{-1/2}$. Thus, *all* Lorentz boosts and spatial rotations are just choices for the Λ matrix subject to the restriction (20).

The standard Lorentz-covariant equations of vacuum electrodynamics involving field strengths are easily obtained from Equation (19) upon using the standard definition $F_{\mu\nu} \equiv \partial_\mu A_{P\nu} - \partial_\nu A_{P\mu}$, leading immediately to the transformation law under the Λ-transformations: $F^{(\Lambda)}_{\mu\nu} = \Lambda^\rho_\mu \Lambda^\sigma_\nu F_{\rho\sigma}$, and the equations

$$\begin{aligned} \partial_\mu F^{\mu\nu} &= -J^\nu, \\ \partial_\mu(e^{\mu\nu\rho\sigma} F_{\rho\sigma}) &= 0. \end{aligned} \tag{21}$$

We acknowledge the influence of the Feynman lectures on physics [5] in basing the formulation presented above on the potentials rather than the fields. However, the delineation of the central role of the *projection operator* inherent in the Maxwell equations, for stationary electromagnetism above, as also for the full theory, to be given in the subsection to follow, is original to the best of our knowledge.

3.2. Closing the Circle: Physical Vector Potentials from the Standard Formulation

3.2.1. With Sources

We begin by defining the field strengths $F_{\mu\nu}$ in terms of the standard *gauge potential* A_μ (i.e., without the subscript 'P'): $F_{\mu\nu} \equiv \partial_\mu A_\nu - \partial_\mu A_\nu$. The second of the standard Maxwell equations (21) results immediately. The first of (21) is then either postulated on the basis of experiments, or derived from the Maxwell action [3]. Be that as it may, one may substitute the above definition of the field strengths $F_{\mu\nu}$ in terms of the gauge potentials, yielding

$$\partial_\mu \partial^\mu A^\nu - \partial^\nu \partial_\mu A^\mu = -J^\nu, \tag{22}$$

which, under Fourier transformation (in four dimensions) with Fourier variable k^i, $i = 0, 1, 2, 3$, leads to the equation

$$-k^2 \mathcal{P}^\nu_\mu \tilde{A}(k)^\mu = \tilde{J}^\nu(k). \tag{23}$$

$$\mathcal{P}^\nu_\mu \equiv \delta^\nu_\mu + \frac{k^\nu k_\mu}{k^2}, \quad k^2 \equiv k_\rho k^\rho \neq 0. \tag{24}$$

We first confine to the inhomogeneous Maxwell equation, and take up the homogeneous case later. The projection operator $\mathcal{P}^\nu_\mu$ above possesses the properties characteristic of projection operators in general.

$$\mathcal{P}^\mu_\nu \mathcal{P}^\nu_\rho = \mathcal{P}^\mu_\rho. \tag{25}$$

$$\mathcal{P}^\mu_\nu k^\nu \tilde{f}(k) = 0 \, \forall \, \tilde{f}(k). \tag{26}$$

where (26) characterizes the vectors in the kernel of the projection operator.

The fact that the vacuum Maxwell equation with sources is expressed uniquely and naturally in terms of a *projection* of the gauge potential, without having to make any choices, is of crucial importance, since the projection is clearly on the gauge-invariant physical subspace. This projected vector potential, defined as $A_{P\mu} \equiv \mathcal{P}^\nu_\mu A_\nu$, has the following essential properties, which can be easily gleaned from Fourier space: (a) $\partial_\mu A^\mu_P = 0$, i.e., it is spacetime transverse; (b) under gauge transformations $A_\mu \to A^{(\omega)}_\mu = A_\mu + \partial_\mu \omega$, the projected (physical) vector potential $A^{(\omega)}_{P\mu} = A_{P\mu}$, i.e., it is gauge-invariant and hence, *physical*! This implies that

$$A_\mu = A_{P\mu} + \partial_\mu a, \tag{27}$$

so that the entire burden of gauge transformations of A_i is carried by $a(x) : A_\mu \to A_\mu + \partial_\mu \omega \Rightarrow a^{(\omega)} = a + \omega$, which underlines the complete unphysicality of the pure gauge part ('longitudinal' mode) a of A. It also follows trivially that $F_{\mu\nu}(A) = F_{\mu\nu}(A_P)$, which means that invariance under gauge transformations does not represent a physical symmetry, but merely a *redundancy* in the gauge potential [6]. One also sees that Equation (19) results immediately from our consideration, so we have come full circle. In fact, in Fourier space, we have an explicit solution for the physical potential $A_{P\mu}$ in terms of the sources:

$$\tilde{A}_{P\mu}(k) = -\frac{\tilde{J}_\mu(k)}{k^2} \tag{28}$$

so that, given the form of the 4-vector source, the physical potential and field strengths are determined in spacetime through appropriate inverse Fourier transforms.

In our proof of the gauge invariance of the projected 4-vector potential $\mathbf{A}_P$ above, the special case of gauge functions ω satisfying $\Box^2 \omega = 0$ has been excluded. We notice that if such gauge functions are retained, the projected 4-vector potential is seen have a *residual* gauge ambiguity involving the spacetime gradient of such gauge functions. This residual ambiguity arises even if we impose the Lorentz–Landau gauge as in the standard textbooks. Note that this residual ambiguity may be eliminated, as in standard procedure, if we restrict our attention to electromagnetic field strengths that decay to vanishingly small values at infinity. This implies that the projected physical potential must vanish at infinity as well, leaving the longitudinal mode a to turn into a constant at most at infinity. This implies that the only gauge transformations that are permitted at infinity and solve the wave equation, are constants, whose gradients vanish everywhere. Thus, just as with stationary magnetic fields, the residual gauge ambiguity is no cause for concern.

3.2.2. Without Sources

Consider now the homogeneous or *null* case, when $J^\mu = 0$ in Equation (22). In this case, if $k^2 \neq 0$, the projected vector potential, which is proved to be physical and gauge-invariant above, must vanish, leaving only the unphysical pure gauge part which leads to vanishing field strengths. That solution is devoid of any physical interest.

Thus, it is obvious that for nontrivial electromagnetic fields, $k^2 = 0$, i.e., **k** is a null spacetime vector. In this case, it also follows from the Fourier transformed version of (22) that, for every nontrivial null vector **k**, we must have

$$\mathbf{k} \cdot \mathbf{A} = 0 \Rightarrow \partial_\mu A^\mu = 0. \tag{29}$$

Observe that this is not a choice, but simply follows from the standard Maxwell equations written out in terms of the vector potential.

We recall that 4-dimensional spacetime $\mathcal{M}(3,1)$ can be represented at every point as the Cartesian product $\mathcal{M}(1,1) \times \mathbf{R}^2$, where the first factor is two-dimensional Lorentzian spacetime, and the second is just the Euclidean plane [7]. With this, we realize that there is another null vector **n** linearly independent of **k**, which, together with **k**, spans $\mathcal{M}(1,1)$. One can always choose **n** such that $\mathbf{n} \cdot \mathbf{k} = -1$ for our signature of the Lorentzian metric.

Now, what is known about propagation of electromagnetic waves in vacuum [3,4] is that these waves are endowed with *transverse* spatial polarization, i.e., the electric field (and hence the 3-vector potential) must oscillate in a plane transverse to the spatial direction of propagation. This implies that, for freely propagating electromagnetic waves in vacua, the 4-vector potential cannot possibly have any component in the direction of propagation of light in *spacetime*, thereby precluding any components tangent to $\mathcal{M}(1,1)$. It can have only two *spacelike* physical components, both lying in the Euclidean plane $\mathbf{R}^2$. These physical requirements must be encoded in the projection operator for the source-free case.

It follows that the physical vector potential, defined as

$$A_{P\mu} \equiv \mathcal{P}_{\mu\nu} A^\nu, \quad \mathcal{P}_{\mu\nu} \equiv \eta_{\mu\nu} + k_\mu n_\nu + n_\nu k_\mu, \tag{30}$$

has the properties of being transverse to $\mathcal{M}(1,1)$, having two components both of which are tangential to the Euclidean plane, and is also *gauge-invariant*.

$$\mathbf{k} \cdot \mathbf{A}_P = 0 = \mathbf{n} \cdot \mathbf{A}_P. \tag{31}$$

The latter property follows from the uniqueness of solutions of the two-dimensional Laplace equation. Thus, the gauge-invariant, physical components of the 4-potential satisfying the homogeneous wave equation have their polarization vectors pointing in the two linearly-independent directions of the Euclidean plane $\mathbf{R}_2$. It follows that the projected physical 4-potential A_P is *spacelike* in character. Thus, gauge-invariance is at the root of transverse polarization of electromagnetic waves in vacuum.

4. Applications of the Physical Potentials

4.1. Coupling to Charged Scalar Fields

The action for a self-interacting charged scalar field ψ is, in general, given by the action

$$S_0[\psi] = \int d^4x [\partial_\mu \psi (\partial^\mu \psi)^* - V(|\psi|)]. \tag{32}$$

Using the radial decomposition $\psi = (1/\sqrt{2})\rho(x)\exp i\Theta(x)$, this action is rewritten as

$$S_0[\rho,\Theta] = \int d^4x[\frac{1}{2}\partial_\mu\rho\partial^\mu\rho + \frac{1}{2}\rho^2\partial_\mu\Theta\partial^\mu\Theta - V(\rho)]\,. \tag{33}$$

This action is clearly invariant under the *global* $U(1)$ transformation $\rho \to \rho$, $\Theta \to \Theta + \omega$, for a constant real parameter ω. The conserved Nöther current corresponding to this global symmetry is given by $J^\mu = \rho^2 \partial^\mu\Theta$.

We now add the Maxwell action $S_{Max}[A_P] = -(1/4)\int d^4x F^{\mu\nu}(A_P)F_{\mu\nu}(A_P)$ to S_0. The coupling of the charged scalar field to the physical potential is now affected through the interaction term $S_1 = \int d^4x J_\mu A_P^\mu$, leading to the full action $S[\rho,\Theta,A_P] = S_0 + S_{Max} + S_1$. It is obvious that S is also symmetric under global $U(1)$ transformation of the Θ field, with both ρ, A_P remaining invariant. However, because of the additional interaction term S_1, the conserved Nöther current corresponding to the global symmetry is now augmented to $J'_\mu = \rho^2(\partial_\mu\Theta + A_{P\mu})$. Following the prescription of the iterative Nöther coupling [8], we now couple this augmented current to the physical potential A_P, so as to obtain the full action

$$S'[\rho,\Theta,A_P] = \int d^4x[-\frac{1}{4}F_{\mu\nu}F^{\mu\nu} + \frac{1}{2}\partial_\mu\rho\partial^\mu\rho + \frac{1}{2}\rho^2(D_\mu\Theta D^\mu\Theta) - V(\rho)]\,, \tag{34}$$

where $D_\mu\Theta \equiv \partial_\mu\Theta + A_{P\mu}$. This action is clearly invariant under global $U(1)$ symmetry transformations. Further iterations of the Nöther current interaction leads to no new terms in the action [8].

If this action is rewritten in terms of the *full* gauge potential $A^\mu \equiv A_P^\mu + \partial^\mu a$, with $a(x) \equiv \int d^4x \mathcal{G}(x-x')\partial_\mu A^\mu(x')$, $\Box^2\mathcal{G}(x-x') = \delta^{(4)}(x-x')$, and we introduce a new field $\chi \equiv \Theta - a$, then $D_\mu\Theta = \partial_\mu(\chi + a) + (A_\mu - \partial_\mu a) = \partial_\mu\chi + A_\mu \equiv D_\mu\chi$. Writing $\phi = (1/\sqrt{2})\rho \exp i\chi$, we obtain the net action

$$S'[\phi,A] = \int d^4x[D_\mu\phi(D^\mu\phi)^* - V(|\phi|) - \frac{1}{4}F_{\mu\nu}(A)F^{\mu\nu}(A)], \tag{35}$$

where $D_\mu\phi = \partial_\mu\phi + iA_\mu\phi$. This action is clearly invariant under *local* $U(1)$ gauge transformations: $\phi \to \phi \exp i\omega(x)$, $A_\mu \to A_\mu - \partial_\mu\omega$. Thus, the iterative Nöther coupling prescription leads uniquely to the $U(1)$ gauge-invariant action for the charged scalar field (35). The starting point is of course the physical vector potential $A_{P\mu}$. In terms of the fields ρ, χ, A_P, and a, the local $U(1)$ gauge transformations do not affect ρ, A_P but only $\chi \to \chi + \omega$, $a \to a - \omega$. The minimal coupling prescription is not used here, but emerges from the prescription of iterative Nöther coupling.

4.2. Aharonov–Bohm Effect

The Aharonov–Bohm effect [9] is historically the first tested proposal to underline the physicality of magnetic vector potential $\vec{A}_P$. This effect is a quantum mechanical effect which shows that the wave function of an electron in a closed orbit in a classical vector potential (even if there is no magnetic field in the region) will pick up a geometric phase given by the anholonomy of the vector potential along the closed curve. This phase is thus given by the expression $\Phi(\vec{A}) = \int_C \vec{A}\cdot d\vec{l}$, where C is a noncontractible loop. Now, in our approach, the vector potential admits the decomposition $\vec{A} = \vec{A}_P + \vec{A}_U$, where $\vec{A}_U = \nabla a(\vec{x})$ with $a(\vec{x}) \equiv \int d^3x' \nabla' \cdot \vec{A}(x')/|\vec{x}-\vec{x}'|$. If the scalar $a(\vec{x})$ is single-valued everywhere on C, then it is obvious that $\Phi(\vec{A}) = \Phi(\vec{A}_P)$! In other words, the physically measured geometric phase $\Phi(A_P)$ is dependent only on the physical projection A_P of the gauge potential, and is quite independent of the pure gauge part dependent on $a(x)$.

5. Conclusions

We saw in the last section that physical effects stemming from nontrivial configuration spaces of test charges, like the Aharonov–Bohm effect, actually reinforce our contention that the gauge-invariant projection of the 4-vector potential plays the key role at the expense of the pure gauge piece. This approach completely demystifies the topic of gauge ambiguity, and champions special relativity through a totally Lorentz-covariant approach, free of gauge ambiguities.

Recently, it has been shown [8] that any *non-Abelian* gauge theory (with matter interactions) is classically equivalent to a set of Abelian gauge fields, whose self-interaction and interaction with matter are generated by a process of iterative Nöther coupling, without invoking the minimal coupling prescription. Since Abelian gauge fields are completely described by their physical projection, as elaborated in this paper, a mathematically simpler description of non-Abelian gauge fields, avoiding any Faddeev–Popov gauge fixing, can be envisaged using our results. A preliminary attempt in this direction has been made in [6]. We hope to report more complete results elsewhere.

A related issue is that our approach can avoid the conundrum discussed many years ago by Gupta and Bleuler [10], associated with canonical quantization of the free Maxwell field, when the gauge potential is gauge fixed by means of a Lorentz-invariant gauge condition like the Lorenz–Landau gauge. Due to the indefinite spacetime metric, states in the Fock space of the theory are seen to possess negative norm. Gupta and Bleuler proposed that these unphysical Fock space states must be eliminated by subsidiary conditions imposed on the Fock space. In our approach, the projected 4-potential is actually a *spacelike* 4-vector with vanishing projection along the two linearly independent null directions of Minkowski 4-spacetime. The physical subspace of polarizations is $\mathcal{R}_2$, so problems associated with the indefinite metric of Minkowski spacetime ought not to be of consequence. We hope to report on this in detail elsewhere.

We have also been recently informed that similar projection operators have been considered in some contemporary works on quantum field theory, e.g., [11–14]. Even earlier, it was apparently J. L. Synge who first proposed projection operators to project out the physical degrees of freedom [15] of the electromagnetic field interacting with test charges. However, the formulation given here is that of the authors of this paper. Earlier, it has also been extensively discussed in class lectures on Maxwell electrodynamics given by one of us (PM) since 2005.

Author Contributions: The authors contribute equally to this paper.

Funding: This research received no external funding.

Acknowledgments: One of us (P.M.) would like to thank J. Navarro-Salas for his immense help in submitting this paper to Symmetry and his subsequent advise regarding publication.

Conflicts of Interest: The authors declare no conflict of interest.

References

1. Yang, C.N. The conceptual origins of Maxwell's equations and gauge theory. *Phys. Today* **2014**, *67*, 45. [CrossRef]
2. Einstein, A. Zur Elektrodynamik bewegter Körper. *Ann. Phys.* **1905**, *17*, 37; reprinted as On the Electrodynamics of Moving Bodies. In *The Principle of Relativity*; Dover: New York, NY, USA, 1952. [CrossRef]
3. Landau, L.D.; Lifschitz, E.M. *Classical Theory of Fields*; Pergamon Press: Oxford, UK, 1975.
4. Jackson, J.D. *Classical Electrodynamics*; Wiley Eastern: Hoboken, NJ, USA, 1974.
5. Feynman, R.P. *The Feynman Lectures on Physics*; Addison-Wesley Publishing Co.: Boston, MA, USA, 1963; Volume 2.
6. Bhattacharjee, S.; Majumdar, P. Gauge-free Coleman-Weinberg Potential. *Eur. Phys. J. C* **2013**, *73*, 2348; e-Print arXiv:1302.7272.
7. Synge, J.L. Model Universes with Spherical Symmetry. *Ann. Mat. Pura Appl.* **1974**, *86*, 239–255. [CrossRef]
8. Basu, A.; Majumdar, P.; Mitra, I. Gauge-invariant Matter Field Actions from Iterative Nöther Coupling. *Phys. Rev. D* **2018**, *98*, 105018; e-Print arXiv: 1711.05608.
9. Aharonov, Y.; Bohm, D. Significance of Electromagnetic Potentials in the Quantum Theory. *Phys. Rev.* **1959**, *115*, 485. [CrossRef]
10. Itzykson, C.; Zuber, J.B. *Quantum Field Theory*; McGraw-Hill Inc.: New York, NY, USA, 1980.
11. Srednicki, M. *Quantum Field Theory*; Cambridge University Press: Cambridge, UK, 2007.

12. Van Holten, J.W. Aspects of BRST quantization. In *Topology and Geometry in Physics*; Springer: Berlin/Heidelberg, Germany, 2005.
13. Bick, E.; Steffen, F.D. (Eds.) Springer Lecture Notes in Physics 659. Available online: https://www.springer.com/gp/book/9783540231257 (accessed on 10 June 2019).
14. Online Lecture Notes on Quantum Field Theory by David Tong. Available online: http://www.damtp.cam.ac.uk/user/tong/qft.html (accessed on 10 June 2019).
15. Synge, J.L. Point Particles and Energy Tensors in Special Relativity. *Ann. Mat. Pura Appl.* **1970**, *84*, 33–59. [CrossRef]

© 2019 by the authors. Licensee MDPI, Basel, Switzerland. This article is an open access article distributed under the terms and conditions of the Creative Commons Attribution (CC BY) license (http://creativecommons.org/licenses/by/4.0/).

Article

A Non-Local Action for Electrodynamics: Duality Symmetry and the Aharonov-Bohm Effect, Revisited

Joan Bernabeu [1,2] and Jose Navarro-Salas [3,*]

[1] Physik Department, Ludwig-Maximilians-Universität München, Theresienstraße 37, D-80333 München, Germany; Joan.Bernabeu@physik.uni-muenchen.de
[2] Physik Department, Technische Universität München, James Franck Straße 1, D-85748 Garching, Germany
[3] Departamento de Física Teórica and IFIC, Centro Mixto Universidad de Valencia-CSIC, Facultad de Física, Universidad de Valencia, Valencia, 46100 Burjassot, Spain
* Correspondence: jnavarro@ific.uv.es

Received: 5 September 2019; Accepted: 19 September 2019; Published: 21 September 2019

Abstract: A non-local action functional for electrodynamics depending on the electric and magnetic fields, instead of potentials, has been proposed in the literature. In this work we elaborate and improve this proposal. We also use this formalism to confront the electric-magnetic duality symmetry of the electromagnetic field and the Aharonov–Bohm effect, two subtle aspects of electrodynamics that we examine in a novel way. We show how the former can be derived from the simple harmonic oscillator character of vacuum electrodynamics, while also demonstrating how the magnetic version of the latter naturally arises in an explicitly non-local manner.

Keywords: non-local action; electrodynamics; electromagnetic duality symmetry; Aharonov-Bohm effect

1. Introduction

Locality is a preferred virtue of fundamental field theories. Electrodynamics, the paradigm of field theory, and general relativity, the modern and finest description of gravity, are very important examples. Both theories are consistent with local causality and the conservation of energy and momentum. Maxwell's and Einstein's equations are systems of partial differential equations for their fundamental fields: the electromagnetic and metric tensors, respectively. The two sets of field equations can also be derived from an action functional. The Hilbert-Einstein action itself is also local in the metric field. However, to derive the Maxwell equations from a local action one has to introduce the electromagnetic potentials. To construct an action depending exclusively on gauge invariant quantities one must necessarily sacrifice locality. This issue is very rarely treated in the literature, despite of the fact that it is a question that may naturally arise in graduate courses on basic field theory and classical electrodynamics (see, for instance [1,2] and references therein). Within the context of constrained dynamical systems [3–5], a non-local action functional describing Maxwell theory, dependent on the electric and magnetic fields, was sketched in Ref. [6]. In this paper we will focus on this proposal and related aspects of quantum mechanics and the theory of Noether's symmetries.

As remarked above, electrodynamics is commonly formulated in terms of Hamilton's variational principle through the action functional $S[A^\mu] = \int d^4 x \mathcal{L}_{\text{EM}}$, where the Lagrangian density for the electromagnetic field in the presence of an external current source $J^\mu \equiv (\rho, \mathbf{J})$, is given by [7,8]

$$\mathcal{L}_{\text{EM}} \equiv -\frac{1}{4} F_{\mu\nu} F^{\mu\nu} - A^\mu J_\mu \,. \tag{1}$$

The action is regarded as a functional of the 4-vector potential $A^\mu = (A^0, \mathbf{A})$, where $F^{\mu\nu} \equiv \partial^\mu A^\nu - \partial^\nu A^\mu$ is the electromagnetic field tensor. $E^i = -F^{0i}$ and $B^i = -\frac{1}{2}\epsilon^{ijk} F^{jk}$ are the components of the electric and magnetic fields ($\mathbf{E}$ and $\mathbf{B}$), respectively, and the metric $\eta = \text{diag}(1, -1, -1, -1)$ was used to lower and raise indices in J_μ, $F_{\mu\nu}$, and ∂^μ (e.g., $J_\mu = \eta_{\mu\nu} J^\nu$). [Throughout this work we use Lorentz-Heaviside units and take $c = 1$. We also assume the Einstein summation convention for repeated indices and $\epsilon^{123} = 1$. Additionally, greek letter indices refer to time and Cartesian space coordinates whereas latin letter indices only refer to the latter. Furthermore, simultaneous spacetime points are labelled as $x \equiv (t, \mathbf{x})$ and $x' \equiv (t, \mathbf{x}')$. Finally, it is assumed that all fields decay to 0 at infinity.]

The inhomogeneous Maxwell equations

$$\nabla \times \mathbf{B} - \partial_t \mathbf{E} = \mathbf{J}, \tag{2}$$

$$\nabla \cdot \mathbf{E} = \rho, \tag{3}$$

are obtained by varying the action with respect to δA^μ and imposing $\delta S = 0$. One gets immediately $\partial_\mu (\partial^\mu A^\nu - \partial^\nu A^\mu) = J^\nu$, and rewriting the potential in terms of the electric and magnetic fields, Gauss' law (3) and the Ampere-Maxwell equation (2) are readily obtained. The fact that (2) and (3) only hold on-shell (i.e., when the Euler-Lagrange equations for A^μ hold) contrasts with the off-shell nature of the homogeneous Maxwell equations

$$\nabla \times \mathbf{E} + \partial_t \mathbf{B} = 0, \tag{4}$$

$$\nabla \cdot \mathbf{B} = 0, \tag{5}$$

which are trivially satisfied by the definition of $F^{\mu\nu}$ in terms of the potentials, or equivalently $\mathbf{E} = -\nabla A^0 - \frac{\partial}{\partial t}\mathbf{A}$, $\mathbf{B} = \nabla \times \mathbf{A}$ in vector notation. This distinction between two types of Maxwell equations can seem somewhat forced, as in essence it is only due to the choice of A^μ as the field of the action functional. Nevertheless, it is the price to be paid to deal with a *local* action, i.e., one where $\mathcal{L}_{\text{EM}}$ depends on the value of $A^\mu(x)$ and finitely many derivatives at a single spacetime point x.

An alternative local action functional is given by [2]

$$S[A_\mu, F_{\mu\nu}] = \int d^4 x \left[\frac{1}{4} F_{\mu\nu} F^{\mu\nu} - \frac{1}{2} F^{\mu\nu}(\partial_\mu A_\nu - \partial_\nu A_\mu) - A^\mu J_\mu \right]. \tag{6}$$

$F_{\mu\nu}$ and A_μ are here considered to be completely independent dynamical variables. The equation of motion for $F_{\mu\nu}$ is $F_{\mu\nu} = \partial_\mu A_\nu - \partial_\nu A_\mu$, and plugging this into the action (6) one gets the standard action $S[A^\mu] = \int d^4 x \mathcal{L}_{\text{EM}}$. This alternative first-order action (6) is very efficient to prove [2] that the covariant Feynman rules for quantum electrodynamics obtained from the functional integral approach are indeed equivalent to the rules derived within the canonical formalism.

The use of potentials in (1) is also useful to study electrodynamics with matter sources. Recycling the field-matter interaction term $-A^\mu J_\mu$ present in (1), inserting the charge distribution (the dot refers to a total time derivative)

$$\rho(x') = e\delta^3(\mathbf{x}(t) - \mathbf{x}') \quad \text{and} \quad \mathbf{J}(x') = e\dot{\mathbf{x}}(t)\delta^3(\mathbf{x}(t) - \mathbf{x}'), \tag{7}$$

and adding a kinetic energy term, the standard Lagrangian that describes the motion of a non-relativistic particle of mass m and charge e within an external electromagnetic field,

$$L_p = \frac{1}{2} m \dot{\mathbf{x}}^2 + e\mathbf{A} \cdot \dot{\mathbf{x}} - eA^0, \tag{8}$$

is recovered. Despite the fact that the action $S_P[\mathbf{x}] = \int dt L_P$ is explicitly dependent on the potentials, the equations of motion, which in this case are just the Lorentz force

$$m\ddot{\mathbf{x}} = e(\mathbf{E} + \dot{\mathbf{x}} \times \mathbf{B}), \qquad (9)$$

can be expressed solely in terms of the electromagnetic field, similarly to the case of Equations (2) and (3) with respect to the action S. Consequently, in classical mechanics where $\delta S_P = 0$ strictly defines the dynamics of the particle, this formulation does not pose anything more than possibly an aesthetic nuisance. However, in the context of quantum mechanics, where the contribution of trajectories with $\delta S_P \neq 0$ to the path integral is not negligible [9], this formulation does become an issue with the interpretation of the Aharonov–Bohm (AB) effect [10–14].

As mentioned above, the first aim of this paper is to study the non-local formulation suggested by Jackiw [6]. It is of first-order in time derivatives, but spatially non-local. We will elaborate on this proposal finding a slightly more simplified expression for the action functional than that originally proposed [6] (see the comments after Equation (29)). This alternative non-local action turns out to be very efficient to analyze the electric-magnetic duality symmetry of free electrodynamics, and, as a bonus, to gain new insights on the AB effect.

2. The Free Non-Local (Duality Invariant) Action

A wide family of first-order Lagrangians in classical mechanics can be expressed as

$$L = \omega_{ij} \dot{q}^i p^j - H(q, p), \qquad (10)$$

where the constants ω_{ij} are the components of the off-diagonal block term of the symplectic tensor

$$\Omega = \begin{pmatrix} 0 & \omega \\ -\omega & 0 \end{pmatrix} \qquad (11)$$

and $H(q, p)$ is the system's Hamiltonian [3,6]. As the notation hints, $q = \{q^i\}$ and $p = \{p^i\}$ are the sets of (phase space) variables. If ω has an inverse ω^{-1}, then their brackets are simply $\{q^i, p^j\} \equiv \omega^{ij}$ ($\{q^i, q^j\} = \{p^i, p^j\} = 0$), where ω^{ij} are the components of ω^{-1}. The conventional choice for simple Hamiltonian systems is $\omega_{ij} = \delta_{ij}$, and hence q and p are canonically conjugate variables with $\{q^i, p^j\} = \delta^{ij}$. However, when ω is not invertible, one typically faces a constrained system, examples of which we give below.

The Lagrangian (10) can be generalized to a Lagrangian density for the context of field theory. Besides summing over the discrete degrees of freedom in the non-Hamiltonian component of (10), one must also sum over (i.e., integrate) the continuous degrees of freedom. Thus, the Lagrangian density of the conjugate fields ϕ and π can be expressed in terms of the Hamiltonian density $\mathcal{H}(\phi, \pi)$ as

$$\mathcal{L} = \int d^3x' \omega_{ij}(\mathbf{x}, \mathbf{x}') \partial_t \phi^i(x) \pi^j(x') - \mathcal{H}(\phi, \pi) \qquad (12)$$

with $\{\phi^i(x), \pi^j(x')\} \equiv \omega^{ij}(\mathbf{x}, \mathbf{x}')$, if ω is invertible. The most conventional choice for ω in field theory is $\omega_{ij}(\mathbf{x}, \mathbf{x}') = \delta_{ij} \delta^3(\mathbf{x} - \mathbf{x}')$, which leads to the local Lagrangian density $\mathcal{L} = \partial_t \phi^i(x) \pi^j(x) - \mathcal{H}(\phi, \pi)$. For

$$\mathcal{H}(\phi, \pi) = \frac{1}{2}(\pi^2 + (\nabla \phi)^2) + \frac{1}{2} m^2 \phi^2 \qquad (13)$$

we have the usual free scalar Klein-Gordon theory, with field equations $\partial_t \phi = \pi$ and $\partial_t \pi = (\nabla^2 - m^2)\phi$, which easily combine into the Klein-Gordon wave equation $(\partial_t^2 - \nabla^2 + m^2)\phi = 0$, consistent with $\{\phi^i(x), \pi^j(x')\} = \delta^{ij} \delta^3(\mathbf{x} - \mathbf{x}')$.

A more involved example is given by taking $\omega_{ij}(\mathbf{x}, \mathbf{x}')$ as the divergenceless or transverse delta function

$$\omega_{ij}(\mathbf{x}, \mathbf{x}') = \delta_{ij}^{\mathrm{T}}(\mathbf{x} - \mathbf{x}') \equiv \delta_{ij}\delta^3(\mathbf{x} - \mathbf{x}') + \partial_i \partial_j \frac{1}{4\pi |\mathbf{x} - \mathbf{x}'|}. \tag{14}$$

It is convenient to briefly recall here that a generic vector field $\mathbf{F}$ always decomposes univocally [15] into a transverse vector $\mathbf{F}_\mathrm{T}$, obeying $\nabla \cdot \mathbf{F}_\mathrm{T} = 0$, plus a longitudinal one $\mathbf{F}_\mathrm{L}$, with $\nabla \times \mathbf{F}_\mathrm{L} = 0$. The transverse delta can then be used to project the transverse component,

$$\int d^3 x' \delta_{ij}^{\mathrm{T}}(\mathbf{x} - \mathbf{x}') F^j(\mathbf{x}') = F_\mathrm{T}^i(\mathbf{x}). \tag{15}$$

Choosing the variables to be vector fields $\phi \to \mathbf{E}$, $\pi \to \mathbf{A}$ with a Hamiltonian density given by

$$\mathcal{H}_0(\mathbf{E}, \mathbf{A}) = \frac{1}{2}[(\mathbf{E}^2 + (\nabla \times \mathbf{A})^2], \tag{16}$$

then the (non-local) Lagrangian density reads

$$\mathcal{L}_0 = \int d^3 x' \delta_{ij}^{\mathrm{T}}(\mathbf{x} - \mathbf{x}') \partial_t E^i(x) A^j(x') - \mathcal{H}_0(\mathbf{E}, \mathbf{A}). \tag{17}$$

In contrast with the Klein-Gordon example, this Lagrangian density, due to the extra contribution to the delta function, cannot be reduced to a local one in terms of the chosen fields $\mathbf{E}, \mathbf{A}$. Furthermore, Equation (17) is invariant under gauge transformations $\mathbf{A}' = \mathbf{A} + \nabla \xi$. By taking variations and assuming the appropriate boundary conditions one obtains the field equations

$$E^i = -\int d^3 x' \delta_{ij}^{\mathrm{T}}(\mathbf{x} - \mathbf{x}') \partial_t A^j(x') = -\partial_t A_\mathrm{T}^i, \tag{18}$$

$$[\nabla \times (\nabla \times \mathbf{A})]^i = \int d^3 x' \delta_{ij}^{\mathrm{T}}(\mathbf{x} - \mathbf{x}') \partial_t E^j(x') = \partial_t E_\mathrm{T}^i. \tag{19}$$

However, after some manipulations one can transform the above equations into the following set of local field equations

$$\partial_t \mathbf{E} = \nabla \times (\nabla \times \mathbf{A}), \quad \nabla \times \mathbf{E} + \partial_t(\nabla \times \mathbf{A}) = 0, \tag{20}$$

$$\nabla \cdot \mathbf{E} = 0. \tag{21}$$

The source-free versions of (2)–(4) are recovered with the identification $\mathbf{B} = \nabla \times \mathbf{A}$. Equation (5) identically follows from the definition of the magnetic field in terms of $\mathbf{A}$, hence completing the full set of vacuum Maxwell equations. Note how the Gauss law constraint (21) was obtained without explicitly introducing any Lagrange multiplier. Also note how the transverse delta can project $\mathbf{A}_\mathrm{T}$, leading to the Lagrangian density

$$\mathcal{L}_0 = \partial_t \mathbf{E} \cdot \mathbf{A}_\mathrm{T} - \frac{1}{2}\left[\mathbf{E}^2 + (\nabla \times \mathbf{A}_\mathrm{T})^2\right], \tag{22}$$

where the longitudinal component of $\mathbf{A}$ has naturally decoupled from the theory. That this is the case seems natural, as $\mathbf{A}_\mathrm{L}$ does not possess indispensable physical value due to the aforementioned gauge invariance. Please note that although (22) is apparently a local expression, there is a hidden non-locality in the (constrained and gauge-independent) transverse vector potential. Solving now the constraint (21) (i.e., taking $\mathbf{E} = \mathbf{E}_\mathrm{T}$) into (23) we finally get

$$\mathcal{L}_0 = \partial_t \mathbf{E}_\mathrm{T} \cdot \mathbf{A}_\mathrm{T} - \frac{1}{2}[\mathbf{E}_\mathrm{T}^2 + (\nabla \times \mathbf{A}_\mathrm{T})^2]. \tag{23}$$

In this way we therefore recover the completely reduced form of the electromagnetic Lagrangian density. A bonus of the above discussion is that one can immediately work out the brackets of the theory: $\delta_{ij}^{\rm T}(\mathbf{x}-\mathbf{x}')$ can be inverted for transverse vector fields and hence the expected [16–18] $\{E_{\rm T}^i(x), A_{\rm T}^j(x')\} = \delta^{{\rm T}ij}(\mathbf{x}-\mathbf{x}')$ is derived.

2.1. Non-Local Formulation for the Electromagnetic Field in Terms of **E** *and* **B**

Our last and most important example consists of defining the object $\omega_{ij}(\mathbf{x},\mathbf{x}')$ for the electric and magnetic field themselves. The solution involves a derivative of the Green's function for the Laplacian operator $\triangle \equiv \partial_i \partial_i$, and it is given by

$$\omega_{ij}(\mathbf{x},\mathbf{x}') = \epsilon^{ijk}\partial_k \frac{-1}{4\pi|\mathbf{x}-\mathbf{x}'|} \,. \tag{24}$$

This expression can be regarded as the simplest way to enforce the appropriate physical dimensions for $\omega_{ij}(\mathbf{x},\mathbf{x}')\partial_t E^i B^j$ and consistency with respect to electric-magnetic duality symmetry (see next subsection for more details). Together with the conventional electromagnetic Hamiltonian density we can construct, in the absence of sources, the action $S_{\rm NL,0}[\mathbf{E},\mathbf{B}] = \int d^4 x \mathcal{L}_{\rm NL,0}$, a functional exclusively dependent on the electromagnetic field, with a first-order Lagrangian density

$$\mathcal{L}_{\rm NL,0} = \int d^3 x'\, \omega_{ij}(\mathbf{x},\mathbf{x}')\partial_t E^i(x) B^j(x') - \frac{1}{2}(\mathbf{E}^2(x) + \mathbf{B}^2(x)) \,. \tag{25}$$

It is quite remarkable that this action yields all of the four vacuum Maxwell equations. The integral term in (25) introduces an explicit non-locality, as the fields at spatially separated points $x = (t,\mathbf{x})$ and $x' = (t,\mathbf{x}')$ "interact" with one another. This coupling is nonetheless weighed by $\omega_{ij}(\mathbf{x},\mathbf{x}')$, leading it to steadily decay as $\mathbf{x}$ and $\mathbf{x}'$ become further apart. Taking variations of E^i and B^i, simultaneously exploiting the standard fall-off conditions of the fields at infinity, one can show that the equations of motion are just the Hemholtz decomposition [15] of the free electromagnetic field,

$$E^i(x) = -\int d^3 x'\, \omega_{ij}(\mathbf{x},\mathbf{x}')\partial_t B^j(x') \,, \tag{26}$$

$$B^i(x) = \int d^3 x'\, \omega_{ij}(\mathbf{x},\mathbf{x}')\partial_t E^j(x') \,. \tag{27}$$

Applying a divergence and a curl on (26) and (27) immediately provides the vacuum versions of Equations (2)–(5),

$$\nabla \times \mathbf{E} = -\partial_t \mathbf{B} \,, \quad \nabla \times \mathbf{B} = \partial_t \mathbf{E} \,, \tag{28}$$

$$\nabla \cdot \mathbf{E} = 0 \,, \quad \nabla \cdot \mathbf{B} = 0 \,. \tag{29}$$

The non-local Lagrangian density $\mathcal{L}_{\rm NL,0}$ is similar to the one given in Ref. [6], up to the contributions of two Lagrange multipliers, which we find unnecessary in the absence of sources. As in the previous case [(17) and (23)], the constraints (29) can be solved into the Lagrangian density (25). In this situation, where the fields are necessarily transverse, ω does possess an inverse, leading to the anticipated [17] brackets

$$\{E_{\rm T}^i(x), B_{\rm T}^j(x')\} = -\epsilon^{ijk}\partial_k \delta^3(\mathbf{x}-\mathbf{x}') \,. \tag{30}$$

Note also how (17), and consequently (23), can also be recovered from (25) by introducing the vector potential **A** such that $\mathbf{B} = \nabla \times \mathbf{A}$.

2.2. Electric-Magnetic Duality Symmetry

The fact that (25) is formulated solely in terms of **E** and **B** means that it is manifestly dual, quite in contrast to the standard formulation (1). It is straightforward to prove that the discrete transformations

$\mathbf{E} \to -\mathbf{B}$, $\mathbf{B} \to \mathbf{E}$ and their continuous generalization as electric-magnetic duality rotations [7] with parameter θ,

$$\begin{pmatrix} \mathbf{E}' \\ \mathbf{B}' \end{pmatrix} = \begin{pmatrix} \cos\theta & \sin\theta \\ -\sin\theta & \cos\theta \end{pmatrix} \begin{pmatrix} \mathbf{E} \\ \mathbf{B} \end{pmatrix}, \qquad (31)$$

leave the Maxwell equations invariant. It is, however, not such a simple task [19–22] to prove that (31) are a symmetry in the Noether sense, i.e., that their infinitesimal version

$$\delta \mathbf{E} = \theta \mathbf{B}, \qquad \delta \mathbf{B} = -\theta \mathbf{E}, \qquad (32)$$

leaves the Lagrangian $L = \int d^3 \mathcal{L}$ invariant, up to a total time derivative and without making use of the field equations.

Employing the standard formulation (1), the transformations (32) clearly will not suffice as Noether's theorem requires the transformations of the dynamic fields, A^μ in this case. However, the problem is actually deeper. The introduction of the potentials implies that Equations (4) and (5) hold, which for consistency would also require, through the use of (32), the equations $\nabla \times \mathbf{B} - \partial_t \mathbf{E} = 0$ and $\nabla \cdot \mathbf{E} = 0$. However, within the Lagrangian formalism it is forbidden to use the latter (on-shell) equations to prove that the duality rotations are a symmetry of the theory. Consequently, the transformation in (32) cannot be applied directly [20,21] on (1) with Noether's Theorem. A way out of this tension is to project the original duality rotations on the transverse fields ($\mathbf{E}_T$, $\mathbf{A}_T$) and consider the reduced Lagrangian (23) [20,21]. The new form of the duality symmetry is then non-local.

On the other hand, the application of Noether's theorem with (25) is swift and even elegant. While the bracket has become more intricate in the transition from using $\mathbf{A}$ and $\mathbf{E}$ to $\mathbf{B}$ and $\mathbf{E}$, the Hamiltonian density now has the well known form of the isotropic simple harmonic oscillator (SHO),

$$H(q,p) = \frac{1}{2}(q^2 + p^2) \quad \text{(normalized)}. \qquad (33)$$

The presence of the SHO in this context shouldn't be too surprising, as it is a well-known fact that vacuum electromagnetic field satisfies the wave equations $\partial_\mu \partial^\mu \mathbf{E}$ and $\partial_\mu \partial^\mu \mathbf{B} = 0$, which are just the field version of the SHO equations $\ddot{q}^i + k^2 q^i = 0$ and $\ddot{p}^i + k^2 p^i = 0$. Thus, (25) can be viewed as a the first-order Lagrangian of a SHO with *non-canonical*, i.e., $\{q^i, p^j\} \neq \delta^{ij}$, commutation relations. As with the *canonical*, i.e., $\{q^i, p^j\} = \delta^{ij}$, SHO, this system is also invariant under phase space rotations

$$\begin{pmatrix} q'^i \\ p'^i \end{pmatrix} = \begin{pmatrix} \cos\theta & \sin\theta \\ -\sin\theta & \cos\theta \end{pmatrix} \begin{pmatrix} q^i \\ p^i \end{pmatrix}. \qquad (34)$$

However, while in the canonical case this symmetry implies conservation of energy, the non-trivial case preserves a more general quantity, which using Noether's theorem is straight-forwardly shown to be

$$Q = \frac{1}{2}\omega_{ij}(q^i q^j + p^i p^j). \qquad (35)$$

Of course, phase space rotations (34) are just electric-magnetic rotations (31) in the formalism of (25) and (30), where $\mathbf{E}$ and $\mathbf{B}$ are the (non-canonical) dynamic variables. Thus we can conclude that in the context of the non-local formulation exposed here, electric-magnetic duality is analogous to the phase space rotation symmetry of the SHO, with the conserved quantity being

$$Q_D = \frac{1}{2} \iint d^3 x\, d^3 x'\, \omega_{ij}(\mathbf{x}, \mathbf{x}') \left[E^i(x) E^j(x') + B^i(x) B^j(x') \right]. \qquad (36)$$

Assuming now that the electric and magnetic fields are transverse, the vector potentials $\mathbf{A}(x)$ and $\mathbf{Z}(x)$ can be introduced such that $\mathbf{E} = -\nabla \times \mathbf{Z}$ and $\mathbf{B} = \nabla \times \mathbf{A}$. It is then easily proven that the above non-local quantity (36) becomes the local

$$Q_D = \frac{1}{2} \int d^3x [\mathbf{Z} \cdot (\nabla \times \mathbf{Z}) + \mathbf{A} \cdot (\nabla \times \mathbf{A})], \tag{37}$$

equivalent to the conserved charge obtained by Calkin [19] and Deser-Teitelboim [20]. An extended discussion in the quantum theory is given in [22–25].

We would like to remark that the conservation law $\frac{d}{dt} Q_D = 0$ should be modified in the presence of charged matter, since duality rotations are no longer symmetries of the theory. Note that this is somewhat similar to the chirality transformation of fermions [8]. Chirality rotations are symmetries for massless fermions, implying that $\partial_\mu j_5^\mu = 0$, where $j_5^\mu \equiv \bar{\psi} \gamma^\mu \gamma^5 \psi$, and the corresponding conservation of the chiral charge $Q_5 \equiv \int d^3x j_5^0$. In presence of a mass term, $\frac{d}{dt} Q_5 = 0$ would also be modified accordingly.

3. The Non-Local Action with Matter

The non-local action presented in the previous section can be straightforwardly generalized to accommodate for the presence of matter. This is a important issue since the interaction of the electromagnetic field with matter has both fundamental and applied significance. This new action functional $S_{NL}[\mathbf{E}, \mathbf{B}, \lambda] = \int d^4x \mathcal{L}_{NL}$, essentially based on Ref. [6], has the electric and magnetic fields as its dynamical fields as well as a Lagrange multiplier λ that imposes Gauss' law (3) as a constraint,

$$\mathcal{L}_{NL} = \int d^3x' \, \omega_{ij}(\mathbf{x}, \mathbf{x}')[\partial_t E^i(x) + J^i(x)] B^j(x') - \frac{1}{2}(\mathbf{E}^2 + \mathbf{B}^2) - \lambda(\nabla \cdot \mathbf{E} - \rho). \tag{38}$$

In the above expression $\omega_{ij}(\mathbf{x}, \mathbf{x}')$ is again given by (24). We note that a single Lagrange multiplier λ is introduced here, instead of the two employed in Ref. [6]. This Lagrangian provides all four of Maxwell's equations if there is electric charge conservation, i.e., $\dot{\rho} + \nabla \cdot \mathbf{J} = 0$, a prerequisite that is used in the standard formulation (1) as well to preserve gauge invariance. For instance, if the matter field is given by a Dirac spinor ψ, with electric charge q and mass m, we should replace $\rho = q\bar{\psi}\psi$ and $J^i = q\bar{\psi}\gamma^i\psi$ in (38). One can then complete the action by adding the standard local free action for the Dirac field such that the Lagrangian of the complete theory reads

$$\mathcal{L} = (i\bar{\psi}\gamma^\mu \partial_\mu \psi - m\bar{\psi}\psi) + \int d^3x' \, \omega_{ij}(\mathbf{x}, \mathbf{x}')[\partial_t E^i(x) + q\bar{\psi}\gamma^i\psi(x)]B^j(x') - \frac{1}{2}(\mathbf{E}^2 + \mathbf{B}^2) - \lambda(\nabla \cdot \mathbf{E} - q\bar{\psi}\psi). \tag{39}$$

In addition to the constraint (3) enforced by λ, the equations of motion for the action (38) are

$$E^i = \partial_i \lambda - \int d^3x' \omega_{ij}(\mathbf{x}, \mathbf{x}') \partial_t B^j(x') \tag{40}$$

$$B^i = \int d^3x' \omega_{ij}(\mathbf{x}, \mathbf{x}')[\partial_t E^j(x') + J^j(x')], \tag{41}$$

which correspond to the Helmholtz decomposition of the electromagnetic field coupled to an external source. Gauss' law for the magnetic field is recovered by taking the divergence of (41), while the time-dependent Maxwell Equations (2) and (4) are obtained by applying a curl on (41) and (40) respectively.

The standard formalism in terms of the potentials can also be recovered solving the non-time evolving Equation (5). Applying the variable change $\mathbf{B} \to \mathbf{A}$ such that $\mathbf{B} = \nabla \times \mathbf{A}$ along with the relabelling $A^0 \equiv -\lambda$, it can be shown that (38) becomes

$$\mathcal{L} = (\partial_t \mathbf{E} + \mathbf{J}) \cdot \mathbf{A}_T - \frac{1}{2}\left[\mathbf{E}^2 + (\nabla \times \mathbf{A}_T)^2\right] + A^0 (\nabla \cdot \mathbf{E} - \rho), \tag{42}$$

which is of a similar form to (22). Hence, the introduction of the vector potential makes the non-local Lagrangian density become the standard first-order Lagrangian density after removing the excess longitudinal component of **A**. However, it is important to keep in mind that (38) and (42) are not fully equivalent, as the equation $\nabla \cdot \mathbf{B} = 0$ holds as a proper Euler-Lagrange equation for (38), while it is assumed off-shell for (42).

Nevertheless, it is not difficult to see that (38) can be obtained by introducing the explicit expression of $\mathbf{A}_T$ into (42)

$$A_T^i(x) = \int d^3x' \omega_{ij}(\mathbf{x}, \mathbf{x}') B^j(x') \tag{43}$$

and assuming (5) holds. Therefore, even though the formalism in terms of (38) is not equivalent to the one of (1) or (42), in some instances it will be useful to obtain results for the non-local viewpoint by simply substituting (43) wherever **A** appears in results derived from the local viewpoint, which is equivalent to imposing the Coulomb gauge, i.e., $\nabla \cdot \mathbf{A} = 0$ or $\mathbf{A} = \mathbf{A}_T$. This property can be illustrated by considering the Lagrangian of the non-relativistic particle (8). Inserting (43) and relabelling $\lambda \equiv -A^0$, a new Lagrangian is obtained,

$$L_{\mathrm{NL},p}[\mathbf{x}] = \frac{1}{2}m\dot{\mathbf{x}}^2 + e \int d^3x' \omega_{ij}(\mathbf{x}, \mathbf{x}') \, \dot{x}^i B^j(x') + e\lambda(x). \tag{44}$$

Alternatively, (44) could have been obtained by applying the same procedure that was used to obtain (8) on (38). While the Lagrangian $L_{\mathrm{NL},p}$ appears to be non-local with respect to the magnetic field, the equations of motion are expectedly the Lorentz force (9), which is local in both **E** and **B**. This is reassuring, as in a classical $\delta S = 0$ context no possibly non-local phenomenon is observed.

Things are not so simple however in a quantum context, a fact best depicted by considering the magnetic AB effect with Feynman's path integral method. The details of the setup considered here to analyse the AB effect are described in Figure 1. The action for this process is given by $S_{\mathrm{NL},p} = \int dt L_{\mathrm{NL},p}$ with $\lambda = 0$, and it can thus be proven that the propagator for the electrons getting from the source to the screen is

$$K(\mathbf{x}_f, t_f; \mathbf{x}_1, t_1) = \exp\left[\frac{ie}{\hbar}\int_{\text{above}} ds^i \int d^3x' \omega_{ij}(\mathbf{s}, \mathbf{x}') B^j(x')\right] \int_{\text{above}} \mathcal{D}[\mathbf{x}(t)] \exp\left[\frac{iS_0}{\hbar}\right]$$
$$+ \exp\left[\frac{ie}{\hbar}\int_{\text{below}} ds^i \int d^3x' \omega_{ij}(\mathbf{s}, \mathbf{x}') B^j(x')\right] \int_{\text{below}} \mathcal{D}[\mathbf{x}(t)] \exp\left[\frac{iS_0}{\hbar}\right]. \tag{45}$$

This result can be obtained using an analogous method to the one shown in Ref. [14]. The term $S_0 = \int dt \frac{1}{2}m\dot{\mathbf{x}}^2$ is the free particle action while subscripts "above" and "below" in (45) are used to distinguish paths that curl above the cylinder from those that curl below. As it is known from the standard analysis of the AB effect, all paths curling above have a common phase, while those curling below have another, a property that appears explicitly in (45). In contrast to the standard analysis however, these phases are explicitly non-local with respect to the physically relevant quantity, the magnetic field **B** inside the cylinder, instead of being local in the vector potential **A** outside. Therefore, the non-locality suggested by the standard derivation of the magnetic AB effect appears naturally in the non-local prescription of electrodynamics described here. While the result (45) can be derived by simply applying the Coulomb gauge on (8) [26], we stress how here it has really been proven from a more fundamental action (38), and not from an arbitrary choice of gauge.

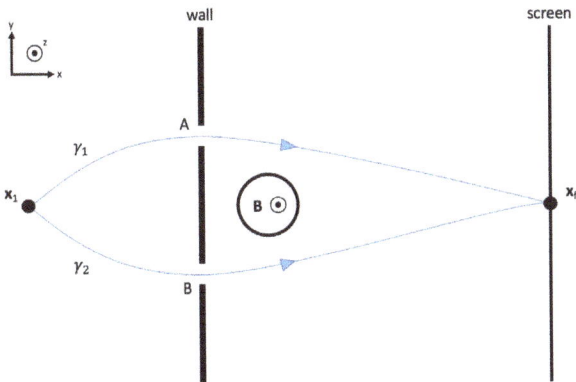

Figure 1. Experimental setup we will consider to analyse the magnetic AB effect. A source of electrons is located at the point $\mathbf{x}_1$, from which one is emitted at a time t_1. Between the source and a screen on the other side of the setup there is a wall, containing two slits A and B, and a long impenetrable cylinder behind it. Inside the cylinder, oriented parallel to the z-axis, there is a magnetic field $\mathbf{B} = \hat{\mathbf{z}} B_0$, while outside $\mathbf{B} = 0$. The electrons can trace two types of *deterministic* paths to reach the point $\mathbf{x}_f$ on the screen at a time t_f, either above (e.g., γ_1) or below (e.g., γ_2) the cylinder.

The cylindrical symmetry of the setup ensures that an analytical value of the nonlocal interaction term, equivalent to the transverse component $\mathbf{A}_T$ of the vector potential (43), can be obtained,

$$\int d^3 x' \omega_{ij}(\mathbf{x}, \mathbf{x}') B^j(x') = \left[\frac{\Phi_\mathbf{B}}{2\pi \rho} \hat{\boldsymbol{\varphi}} \right]^i, \quad (46)$$

where $\rho^2 = (x^1)^2 + (x^2)^2$ is the distance squared with respect to the center of the cylinder and $\hat{\boldsymbol{\varphi}}$ is the unit vector associated with the azimuthal angle. This result can be derived by evaluating the volume integral directly as we have done for completeness in the Appendix A, or treating $\mathbf{A}_T$ as a shorthand for the interaction term (left-hand side (LHS) of (46)) and recycling the standard derivation [14]. The relevant phase difference is thus the expected AB phase,

$$\Delta \varphi = \frac{e}{\hbar} \left[\int_{\text{above}} \mathbf{A}_T \cdot d\mathbf{s} - \int_{\text{below}} \mathbf{A}_T \cdot d\mathbf{s} \right] = \frac{e \Phi_\mathbf{B}}{\hbar}. \quad (47)$$

where $\Phi_\mathbf{B}$ is the magnetic flux through the cylinder.

4. Conclusions

Non-locality is a reasonably objectionable feature, but we feel the fomulation of electrodynamics treated here, elaborating and improving on a proposal sketched in [6], will at least be useful to shed some light on the subtle topic of action functionals independent of potentials. We have argued how non-locality seems to be unescapable in an electromagnetic field-dependent formalism due to the non-trivial commutation relations $\{\mathbf{E}, \mathbf{B}\}$. It is nonetheless important to keep in mind that the field-matter action (38) is not completely independent of potentials, as the Lagrange multiplier λ in (38) is actually just a relabelled (Coulomb gauge) scalar potential. However, it is consistent to assume $\lambda = 0$ in the context of electric-magnetic duality or the magnetic AB effect, meaning they can be studied without concern.

On one hand, the former can be seen as a manifestation of the phase-space rotation symmetry of the SHO. It is worth recalling how this symmetry was derived with an action where all the Maxwell equations hold solely on-shell, in contrast with past derivations, which assume some of them off-shell. On the other hand, an arguably plausible interpretation for the AB effect was deduced. In a classical context, where $\delta S = 0$, the equations of motion (9) of (44) are local in both **E** and **B** despite the non-locality of the action. Therefore the correspondence principle holds, i.e., when $\hbar \to 0$ the interaction of the particle with the electromagnetic field is local. In a quantum context however trajectories with $\delta S \neq 0$ are not negligible, hence the non-locality of the action can materialize (45) with the AB effect. Through this scope, manifest non-locality is thus an exclusively quantum affair, and we believe this is also one of the lessons of this note.

We would like to remark that we are not advocating to avoid the use of field potentials to analyze electrodynamics or its generalizations (nonabelian gauge theories). The purpose of this work is to point out that it could be useful to reanalyze electrodynamics from a nonlocal perspective (using only the electric and magnetic fields). In so doing this we have filled a gap in the literature and obtain, as a bonus, new insights on two important topics in electrodynamics: i) the electromagnetic duality symmetry, and ii) the AB effect.

After finishing this work we became aware of the work [27], concerning a formulation of electrodynamics without a gauge-fixing procedure. We think that there is a close connection with our work that could merit to be further explored.

Author Contributions: The authors contributed equally in writing this article. All authors read and approved the final manuscript.

Funding: This research was funded with Grants. No. FIS2017-84440-C2-1-P, No. FIS2017-91161-EXP, No. SEV-2014-0398, and No. SEJI/2017/042 (Generalitat Valenciana).

Acknowledgments: J.N.-S. thanks I. Agullo and A. del Rio for useful discussions.

Conflicts of Interest: The authors declare no conflict of interest.

Appendix A. Interaction Term in the A.B. Effect

Preliminary considerations:

- The expression for the magnetic field is $\mathbf{B}(\mathbf{x}) = \hat{\mathbf{z}}\Theta(R^2 - x^2 - y^2)$, where Θ is the Heaviside step function and $\mathbf{x} = (x, y, z)$.
- The volume region is a cylinder C of radius R, with a length L_1 and L_2 over and under the xy plane respectively. Furthermore, it will be assumed that the cylinder is long i.e., $L_1^2, L_2^2 \gg R^2, x^2 + y^2$.

Due to its equivalence with the nonlocal interaction term (LHS of (46)), we will use $\mathbf{A}_T$ as a shorthand to refer to it. It can thus be proven that

$$\mathbf{A}_T(\mathbf{x}) = \frac{B_0}{4\pi}\hat{\mathbf{z}} \times \int_C d^3x' \nabla \left(\frac{1}{|\mathbf{x} - \mathbf{x}'|}\right) = \frac{B_0}{4\pi}\hat{\mathbf{z}} \times \int_{\partial C} d\mathbf{S}' \frac{1}{|\mathbf{x} - \mathbf{x}'|}.$$

where a corollary of the Divergence theorem was used in the second equality.

The surface of the cylinder is composed by a circular wall and the two lids on either end. However, since the lids have a normal vector $d\mathbf{S}' \propto \hat{\mathbf{z}}$ and $\hat{\mathbf{z}} \times \hat{\mathbf{z}} = 0$, their contributions to the total integral are 0. Consequently, the only relevant contribution to the integral comes from the circular wall, with a normal vector $d\mathbf{S}' = \hat{\rho}' R d\phi' dz'$ where $\hat{\rho}' = (\cos\phi', \sin\phi', 0)$:

$$= \frac{B_0}{4\pi}\hat{\mathbf{z}} \times \int_0^{2\pi} R d\phi' \hat{\rho}' \int_{-L_2}^{L_1} dz' \left(z'^2 + \alpha(\phi')\right)^{-1/2} \tag{A1}$$

$$= \frac{B_0}{4\pi}\hat{\mathbf{z}} \times \int_0^{2\pi} R d\phi' \hat{\rho}' \left[\log\left(\sqrt{\alpha(\phi') + L_1^2} + L_1\right) + \log\left(\sqrt{\alpha(\phi') + L_2^2} + L_2\right) - \log\left(\alpha(\phi')\right)\right]$$

where $\alpha(\phi') = (x - R\cos\phi')^2 + (y - R\sin\phi')^2$ was introduced for brevity. However, expressions of the form $\log\left(\sqrt{\alpha(\phi') + L^2} + L\right)$ can be disregarded by taking into account the first preliminary consideration,

$$\int_0^{2\pi} d\phi' \hat{\rho}' \log\left(\sqrt{\alpha(\phi') + L^2} + L\right) \approx \log(2L) \int_0^{2\pi} d\phi' \hat{\rho}' = 0.$$

Therefore the expression for $\mathbf{A}_T$ is now a one-dimensional integral

$$\mathbf{A}_T(\mathbf{x}) = -\frac{B_0 R}{4\pi} \int_0^{2\pi} d\phi' \hat{\phi}' \log\left[(x - R\cos\phi')^2 + (y - R\sin\phi')^2\right]. \quad (A2)$$

where $\hat{z} \times \hat{\rho}' = \hat{\phi}'$, with $\hat{\phi}' = (-\sin\phi', \cos\phi', 0)$. Equation (A2) can be reinterpreted as a complex integral, $z_A = A_T^x + iA_T^y$, over a circle of radius R on the complex plane

$$z_A(x,y) = -\frac{B_0}{4\pi} \oint_\gamma dz \log|z - z_0|^2 \quad (A3)$$

where $z_0 = x + iy$ and $\gamma(s) = Re^{is}$. Ignoring for now the multiplicative constant $-B_0/4\pi$, the integral can be split into two,

$$\oint_\gamma dz \log|z - z_0|^2 = \oint_\gamma dz \log(z - z_0) + \oint_\gamma dz \log(z^* - z_0^*) \quad (A4)$$

$$= \oint_\gamma dz \log(z - z_0) + R^2 \oint_\gamma dz \frac{\log(z - z_0^*)}{z^2} \quad (A5)$$

where the latter equality is due to the easily proven general property for circular contour integrals, $\oint_\gamma dz\, f(z^*) = R^2 \oint_\gamma dz\, f(z)/z^2$. The value of (A5) will depend on whether z_0 is inside or outside the disk delimited by γ on the complex plane $\mathbb{C}$ (see Figure A1). In terms of the original problem, this means that the expressions for $\mathbf{A}_T$ inside and outside the solenoid will be different. In the latter case, $\rho^2 \equiv x^2 + y^2 > R^2$, meaning that

$$z_A(x,y) = \frac{B_0 \pi R^2}{2\pi} \frac{-y + ix}{x^2 + y^2}. \quad (A6)$$

and

$$\mathbf{A}_T(\mathbf{x}) = \frac{B_0 \pi R^2}{2\pi(x^2 + y^2)}(-y, x, 0) = \frac{B_0 \pi R^2}{2\pi \rho} \hat{\phi} = \frac{\Phi_B}{2\pi \rho} \hat{\phi} \quad (A7)$$

where $\Phi_B = B_0 \pi R^2$ and $\hat{\phi} = (-y/\rho, x/\rho, 0)$. This is the expected result outside the cylinder. On the other hand, inside $\rho^2 \equiv x^2 + y^2 < R^2$. Evaluating (A5) in this situation gives

$$z_A = \frac{B_0}{2}(-y + ix) \quad (A8)$$

so that

$$\mathbf{A}_T(\mathbf{x}) = \frac{B_0}{2}(-y, x, 0) = \frac{B_0 \rho}{2} \hat{\phi}. \quad (A9)$$

This is the expected result for the transverse component of the vector potential in a finite volume under a constant magnetic field (in this case, a cylinder with $\mathbf{B} = B_0 \hat{z}$), where $\mathbf{A}_T(\mathbf{x}) = -\frac{1}{2}\mathbf{x} \times \mathbf{B}$.

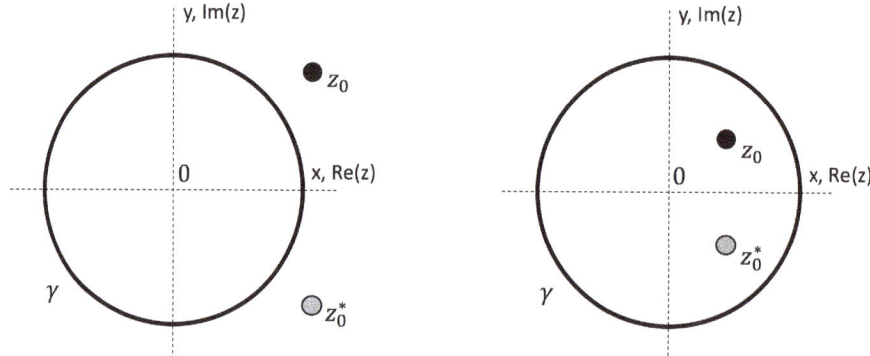

Figure A1. Complex plane representation of $z_0 = x + iy$ outside and inside the cylinder.

References

1. DeWitt, B. Quantum theory without electromagnetic potentials. *Phys. Rev.* **1962**, *125*, 2189. [CrossRef]
2. Coleman, S. *Aspects of Symmetry*; Cambridge University Press: Cambridge, UK, 1985.
3. Faddeev, L.; Jackiw, R. Hamiltonian reduction of unconstrained and constrained systems. *Phys. Rev. Lett.* **1988**, *60*, 1692. [CrossRef] [PubMed]
4. Henneaux, M.; Teitelboim, C. *Quantization of Gauge Systems*; Princeton University: Princeton, NJ, USA, 1992.
5. Barbero, G.F.J.; Díaz, B.; Margalef-Bentabol, J.; Villaseñor, E.J.S. Dirac's algorithm in the presence of boundaries: A practical guide to a geometric approach. *arXiv* **2019**, arXiv:1904.11790.
6. Jackiw, R. Constrained quantization without tears. In *Constraint Theory and Quantization Methods: From Relativistic Particles to Field Theory and General Relativity*; Colomo, F., Lusanna, L., Marmo, G., Eds.; World Scientific: River Edge, NJ, USA, 1994; p. 448, arXiv:hep-th/9306075.
7. Jackson, J.D. *Classical Electrodynamics*; Wiley: New York, NY, USA, 1998.
8. Peskin, M.E.; Schroeder, D.V. *An Introduction to Quantum Field Theory*; Westview Press: Boulder, CO, USA, 1995.
9. Feynman, R.P.; Hibbs, A.R. *Quantum Mechanics and Path Integrals*; McGraw-Hill: New York, NY, USA, 1965.
10. Aharonov, Y.; Bohm, D. Significance of electromagnetic potentials in the quantum theory. *Phys. Rev.* **1959**, *115*, 485. [CrossRef]
11. Aharonov, Y.; Bohm, D. Further considerations on electromagnetic potentials in the quantum theory. *Phys. Rev.* **1961**, *123*, 1511. [CrossRef]
12. Feynman, R.P.; Leighton, R.B.; Sands, M. *The Feynman Lectures on Physics Vol. 2: Mainly Electromagnetism and Matter*; Addison-Wesley: Boston, MA, USA, 1979.
13. Peshkin, M.; Tonomura, A. *The Aharonov-Bohm Effect*; Springer: Berlin, Germany, 1989.
14. Sakurai, J.J.; Napolitano, J. *Modern Quantum Mechanics*; Cambridge University Press: Cambridge, UK, 2017.
15. Griffiths, D.J. *Introduction to Electrodynamics*; Prentice Hall: Upper Saddle River, NJ, USA, 1981.
16. Bjorken, J.D.; Drell, S.D. *Relativistic Quantum Fields*; McGraw-Hill: New York, NY, USA, 1965.
17. Vogel, W.; Welsch, D.G. *Quantum Optics*; John Wiley and Sons: Hoboken, NJ, USA, 2006.
18. Weinberg, S. *Lectures on Quantum Mechanics*; Cambridge University Press: Cambridge, UK, 2015.
19. Calkin, M.G. An invariance property of the free electromagnetic field. *Am. J. Phys.* **1965**, *33*, 958–960. [CrossRef]
20. Deser, S.; Teitelboim, C. Duality transformations of Abelian and non-Abelian gauge fields. *Phys. Rev. D* **1976**, *13*, 1592. [CrossRef]
21. Deser, S. Off-shell electromagnetic duality invariance. *J. Phys. A* **1982**, *15*, 1053. [CrossRef]
22. Agullo, I.; del Rio, A.; Navarro-Salas, J. Electromagnetic duality anomaly in curved spacetimes. *Phys. Rev. Lett.* **2017**, *118*, 111301. [CrossRef]
23. Agullo, I.; del Rio, A.; Navarro-Salas, J. Gravity and handedness of photons. *Int. J. Mod. Phys. D* **2017**, *26*, 1742001. [CrossRef]

24. Agullo, I.; del Rio, A.; Navarro-Salas, J. Classical and quantum aspects of electric-magnetic duality rotations in curved spacetimes. *Phys. Rev. D* **2018**, *98*, 125001. [CrossRef]
25. Agullo, I.; del Rio, A.; Navarro-Salas, J. On the Electric-Magnetic Duality Symmetry: Quantum Anomaly, Optical Helicity, and Particle Creation. *Symmetry* **2018**, *10*, 763. [CrossRef]
26. Stewart, A.M. Role of nonlocality of the vector potential in the Aharonov-Bohm effect. *Can. J. Phys.* **2013**, *91*, 373. [CrossRef]
27. Majumdar, P.; Ray, A. Maxwell Electrodynamics in Terms of Physical Potentials. *Symmetry* **2019**, *11*, 915. [CrossRef]

© 2019 by the authors. Licensee MDPI, Basel, Switzerland. This article is an open access article distributed under the terms and conditions of the Creative Commons Attribution (CC BY) license (http://creativecommons.org/licenses/by/4.0/).

Article

Geometric Objects: A Quality Index to Electromagnetic Energy Transfer Performance in Sustainable Smart Buildings

Juan C. Bravo and Manuel V. Castilla *

Higher Polytechnic School, University of Seville, Seville 41011, Spain; carlos_bravo@us.es
* Correspondence: mviggo@us.es; Tel.: +34-954-556-440

Received: 29 October 2018; Accepted: 28 November 2018; Published: 29 November 2018

Abstract: Sustainable smart buildings play an essential role in terms of more efficient energy. However, these buildings as electric loads are affected by an important distortion in the current and voltage waveforms caused by the increasing proliferation of nonlinear electronic devices. Overall, buildings all around the world consume a significant amount of energy, which is about one-third of the total primary energy resources. Optimization of the power transfer process of such amount of energy is a crucial issue that needs specific tools to integrate energy-efficient behaviour throughout the grid. When nonlinear loads are present, new capable ways of thinking are needed to consider the effects of harmonics and related power components. In this manner, technology innovations are necessary to update the power factor concept to a generalized total or a true one, where different power components involved in it calculation, properly reflect each harmonic interaction. This work addresses an innovative theory that applies the Poynting Vector philosophy via Geometric Algebra to the electromagnetic energy transfer process providing a physical foundation. In this framework, it is possible to analyse and detect the nature of disturbing loads in the exponential growth of new globalized buildings and architectures in our era. This new insight is based on the concept of *geometric objects* with different dimension: vector, bivector, trivector, multivector. Within this paper, these objects are correlated with the electromagnetic quantities responsible for the energy flow supplied to the most common loads in sustainable smart buildings. Besides, it must be considered that these phenomena are characterized by a quality index multivector appropriate even for detecting harmonic sources. A numerical example is used to illustrate the clear capabilities of the suggested index when it applies to industrial loads for optimization of energy control systems and enhance comfort management in smart sustainable buildings.

Keywords: smart building; harmonics; geometric algebra; Poynting Multivector

1. Introduction

1.1. Motivation

Nowadays, professional and academic experts have started to consider the term "Smart Sustainable Cities" [1] so as to incorporate the different aspects of sustainability in the classical "smart cities" new concept. In fact, literature tends to consider a sustainable city as a whole place of sustainable smart buildings that have a strong environmental focus with a balance within the buildings and the city between infrastructures, information and communications technology (ICT), smart technologies, and urban metabolism, focusing mainly on consumption and energy saving [2].

Buildings all around the world consume a significant amount of energy, which is about one-third of the total primary energy resources [3]. For this reason, building energy efficiency has turned out to be a multi-faceted problem and the majority of harmonic problems affecting sustainable smart buildings are generated within new applications and grids of global architectures. This is due, in part, to a proliferation of non-linear loads connected to the networks of the building. These technologies as CCTV recognition systems, automatic smart air conditioning equipment, artificial intelligence (AI), the latest generation of computers, all types of smart detectors and warning systems, and other power electronic equipment are the main sources of such problems. The result of using such highly non-linear load is that the current waveform is distorted. Thus, causing an excessive harmonic of current and voltage. Besides the proximity of many of this new smart building category (industrial and residential smart constructions) will contribute to the distortion of the electric power quality of the feeder, which supplies these constructions and new architectures. These harmonics can cause serious problems in power systems, excessive heat of appliances and machines, aging of electronic component and a decreased capacity, failures of the safety devices and measures of protection, lower power factor and consequently, a reducing power system efficiency due to increasing losses. All these effects are some of the main results of harmonics in power distribution systems. Note, that Harmonic distortion can cause significant costs in distribution networks. Harmonic cost consists of harmonic energy losses, premature aging and de-rating of electrical equipment. The difference between the known generation and the estimated consumption is considered as the energy loss.

Other causes of energy loss and, consequently, an increase in the cost of it, are due to the lack of control of energy efficiency in the thermal performance of buildings and energy balance. Most of the building heat losses occur through the building envelope. In recent years, an important number of papers on quantification and optimization of energy efficiency in buildings has been published referring to the standard for buildings. Many of these works have been developed in different areas of application such energy losses in the building envelope as HVAC systems (heating, ventilation and air conditioning), windows, etc. All these works have in common the aim of making efficient buildings from an energy point of view to be sustainable. Particularly, in [4], an original approach for the U-value evaluation (analogies with coeval buildings, the calculation method, the *in-situ* measurements and the laboratory tests) is taken. In [5], measures of energy efficiency and optimization in the building sector are also evaluated.

In this article, it is proposed an energy quality multivector index (EQI) based on the Generalized Poynting Multivector (GPM) theory, that possesses clear advantages from the viewpoint of harmonic sources detection and minimization in sustainable smart buildings.

1.2. Literature Review

Valuable contributions in electromagnetic field applied to the electric power theory analysis under multi-sinusoidal conditions have been appeared so far in the literature. Despite they have different interpretations, most of them share the common denominator of dealing with the suitability of the Poynting Vector to explain the electromagnetic energy flow in electric systems [6].

In sinusoidal systems, Complex Algebra provides an appropriate framework to analyse the relationship between the Complex Poynting Vector and the energy flow [7].

In one-ports under periodic multi-sinusoidal linear/non-linear operation this issue has still some fundamental unsolved aspects. Nevertheless, some progress has unquestionably been made from numerous valuable contributions in the literature [8–21], each one of them trying to clarify different aspects of the problem by applying the classic Poynting vector (PV) concept. Among them [8,9,14,15] masterfully explain the power factor concept and the physical mechanism of energy propagation in electric power systems; Ferrero et al. in [17] reconsiders the physical meaning assigned to the non-active components of the Park instantaneous power; Balci et al. in [20] describes the transition between PV and instantaneous active and reactive powers; and Faria et al. in [11] computes the instantaneous power directly from Maxwell's equations together with the evaluation of the PV flux. On the other hand, several applications are given by means of PV: Lundin et al. in [10] analyses synchronous generators using field simulations; De León et al. in [12] identifies terms in nonlinear-switched circuits; Cheng et al. in [13] calculates the reactive power of iced transmission line; Todeschini et al. in [16] detects and explains the process of compensation and restoration of symmetry in an unbalanced system and Stahlhut et al. in [21] examines critically the PV possibilities in the area of instrumentation of losses. However, critics of PV calculations [18,19] argue that electromagnetic theory is useless for practical applications of electric power theory. Nevertheless, the large number of papers published on the physical and/or mathematical nature of electromagnetic energy transfer suggest that the work has not been finished.

The multidimensionality of power equation and energy quality indexes in the multi-sinusoidal case is the underlying obstacle that considerably complicates the issue at hand. Instead, reference [13] is a pioneering contribution to the role of a Poynting Multivector to the interpretation of the energy flows based on an original Clifford Vector space. Accordingly, this work adds a new representation of electromagnetic power theory deduced from a Generalized Poynting Multivector [22].

1.3. Contribution

An introduction to "geometric objects" in Geometric Algebra (GA) and the associated phenomena within the electrical systems is developed. In addition, these entities permit a unified treatment of the energy flow concept. By means of the GA framework, the classic explanations of the energy flow process based on interactions between electric and magnetic fields of like-frequency is overcome. The proposed generalization adds the cross-fields interactions in a natural manner, as well.

By this way, this paper is concerned to the application of the Generalized Poynting Multivector (GPM) concept [22] to provide both of them, a physical foundation to the non-active electromagnetic geometric objects as well as a new multivector index to assess the efficiency of the complete energy process in sustainable and smart building loads supplied from a transmission line.

2. Geometric Objects

Geometric Algebra [23–26] is based on the concept of objects with different geometrical dimension that result from the geometric product of distinct graded basis elements, e.g., scalars, vectors, 2-vectors and so on. Thus, starting from vectors within an n-dimensional linear space over the real numbers $\mathcal{V}^n$, the geometric product of vectors ab if $a, b \in \mathcal{V}^n$ is formed by a symmetric inner product

$$a \cdot b = \frac{1}{2}(ab + ba) \tag{1}$$

and an antisymmetric outer product

$$a \wedge b = \frac{1}{2}(ab - ba) \tag{2}$$

Therefore, ab has the canonical decomposition

$$\widetilde{M} = ab = a \cdot b + a \wedge b \tag{3}$$

The resulting multivector $\widetilde{M}$ is the sum of a scalar ($a \cdot b$) and a bivector or 2-vector ($a \wedge b$) object. Despite this sum of two distinct objects might seem strange at first sight and it is against the common rule that only same objects should be added, this Clifford's brilliant idea [23] allows to generalize easily the product to arbitrary higher dimensions and incorporates geometric interpretations of objects and operators.

Thus, a bivector can be viewed as directed plane segment, in much the same way as a vector represents a directed line segment, Figure 1. The bivector $a \wedge b$ has a magnitude $|a \wedge b|$ equal to the usual scalar area of the circle in Figure 1, with the direction of the plane in which the circle lies, and with sense, which can be assigned to the circle in the plane. Then, just as a vector a represents (or is represented by) a directed line segment and a bivector $a \wedge b$ represents a directed plane segment, the trivector (3-vector) $a \wedge b \wedge c$ is a grade-3 object that represents a volume (the sphere).

In $\mathcal{G}_3$, the geometric algebra of three-dimensional space, a general multivector can be written as

$$\widetilde{M} = \alpha + \lambda + B + \mu J \tag{4}$$

where a is a vector, B is a bivector, J is a trivector (pseudoscalar) and λ and μ are both scalars. The three orthogonal basis vectors $\{\sigma_1, \sigma_2, \sigma_3\}$, the three bivectors $\{\sigma_1\sigma_2, \sigma_2\sigma_3, \sigma_3\sigma_1\}$, the unit scalar, and the trivector $\sigma_1\sigma_2\sigma_3$ define a graded linear space of total dimension $8=2^3$ and are shown in Figure 1.

The unit right-handed pseudoscalar for the space J squares to -1, $J^2 = -1$. The pseudoscalar, as well as bivectors, change sign under reversion, but vectors do not. This reversion operation reverses the order of vectors in any product and the convention. So, the reverse of multivector $\widetilde{M}$ derived only from vectors

$$\widetilde{M}^\dagger = ba = a \cdot b - a \wedge b \tag{5}$$

or in the general form resulting from all the possible basis product in $\mathcal{G}_3$

$$\widetilde{M}^\dagger = \alpha + \lambda - B - \mu J \tag{6}$$

For all the formulae presented below the followed notation uses the tilde ~ for multivectors, Euclidean vectors are written in bold font, the dagger symbol † denotes the reverse operation and the upper asterisk * represents the complex conjugate operation. For a complete understanding of notation, a list of symbols is summarized in the glossary.

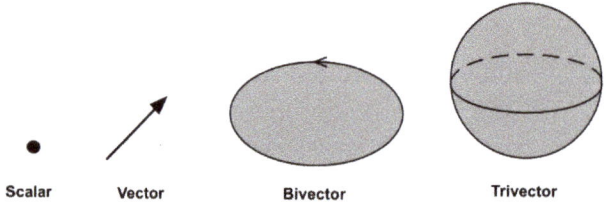

Scalar   Vector   Bivector   Trivector

Figure 1. Geometric objects.

3. The System Model

An "electrical system" is shown in Figure 2 and it can be considered as a space confined by a closed surface to/from which electric power is supplied/received by conductors carrying an electric current. The total instantaneous power transmitted is

$$s(t) = \sum_k u_k i_k \tag{7}$$

Two main electric power processes occur in the system:

- Electric power (energy) is dissipated.
- Electric power (energy) in the system derives from an electric field and magnetic field interactions.

In a general form, the electromagnetic phenomena occurring in the system under study can be roughly modelled by electric circuits.

Despite the application of GA to circuit analysis and power theory has a very short history some relevant advances have been made in this area from different perspectives [27–35].

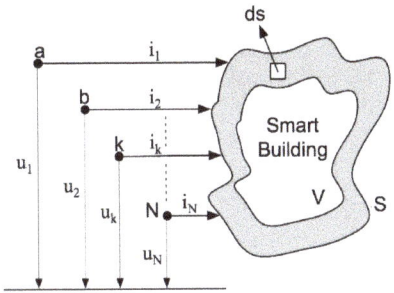

Figure 2. Electrical system.

4. Generalized Poynting Multivector

In this work, space coordinates and time (frequency) coordinates have a differentiated treatment. Space domain remains represented by the classical Euclidean approach, the harmonic field in both, time and frequency domains, are characterized by complex geometric algebra, $\mathcal{CG}_n$. The complete space-time framework results in the hybrid $\mathcal{CG}_n^t - \mathbb{R}^3$ structure named Generalized Euclidean Space ($\mathcal{CG}_n - \mathbb{R}^3$ for space-frequency domain) [22].

In this $\mathcal{CG}_n - \mathbb{R}^3$ algebraic approach, the Generalized Poynting Multivector (GPM) is defined as

$$\widetilde{\Pi} = \sum_{p,q} \left(\widetilde{E}_p \odot \widetilde{H}_q^* \right) \tag{8}$$

where $\widetilde{E}_p$ and $\widetilde{H}_q$ vectors are called 'spatial geometric phasors' of the electric and magnetic harmonic fields, $\widetilde{H}_q^*$ vector is the conjugate of the q-th harmonic spatial geometric phasor $\widetilde{H}_q$ and '$\odot$' is the generalized geometric product (GGP) [27].

From this multivectorial field theory, the flux of the GPM quantity for a volume v enclosed by a surface ς, (GPM theorem), can be expanded into two terms.

$$\iint_\varsigma n \cdot \widetilde{\Pi} \, d\varsigma = \iint_\varsigma \sum_p n \cdot \widetilde{\mathcal{P}}_p \, d\varsigma + \iint_\varsigma \sum_{p \neq q} n \cdot \widetilde{\mathcal{D}}_{pq} \, d\varsigma = \widetilde{S} \tag{9}$$

where n is the unitary vector orthogonal to the infinitesimal surface $d\varsigma$, $\widetilde{\mathcal{P}}$ is the Poynting Multivector (PM) and $\widetilde{\mathcal{D}}$ is the Complementary Poynting Multivector (CPM) and $\widetilde{S} = \widetilde{U} \odot \widetilde{I}^*$ is the Power Multivector.

In Equation (10) are present three "electromagnetic geometric objects": the complex vectors $\widetilde{E}_p$ and $\widetilde{H}_q$, the complex scalar

$$\widetilde{\mathcal{P}} = \sum_p \widetilde{E}_p \odot \widetilde{H}_p^* \tag{10}$$

and the complex bivector

$$\widetilde{\mathcal{D}} = \sum_{p<q} \left(\widetilde{E}_p \odot \widetilde{H}_q^* + \widetilde{E}_q \odot \widetilde{H}_p^* \right) \tag{11}$$

From $\mathcal{CG}_n$ structure, Equation (10) explains the total power energy flow because contains the component $\widetilde{\mathcal{P}}$ related with the power contribution due to frequency-like products and the component $\widetilde{\mathcal{D}}$ associated to the mutual influence of the harmonic components of the fields. Observe that in classical approach this last term vanishes due to the orthogonality of harmonics basis. Nevertheless, a null average value of power without any net energy transferred must be also considered to understand the real power flux processes and to fully evaluate the energetic efficiency of electric systems in presence of harmonics.

The above mentioned electromagnetic geometric objects are shown in Figure 3 and must be considered to explain the main aspects of energy transfer quality.

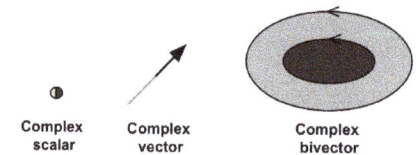

Figure 3. Complex electromagnetic geometric objects.

5. Formulation of Electromagnetic Energy Quality Index

5.1. Energy Flow on Electrical Systems

Consider the case of an electrical system in the form of a circuit, Figure 4, where the load is a linear/nonlinear system. The supply system voltage is periodic but multi-sinusoidal and can be represented by

$$u(t) = \sqrt{2} Im \sum_p U_p e^{j(\omega_p t + \alpha_p)} \quad (12)$$

where p is the harmonic order of $u(t)$. The instantaneous supply current responsible of the generation of the magnetic-field is given by

$$i(t) = \sqrt{2} Im \sum_q I_q e^{j(\omega_q t + \beta_q)} \quad (13)$$

where q is the harmonic order of $i(t)$. It is assumed that a group of voltage harmonics N exist that have corresponding current harmonics of the same frequencies, and that components M of current exist without corresponding voltages. In linear operation, $\beta_q = \alpha_q - \varphi_q$, φ_q is the impedance phase angle of the consumer electrical system.

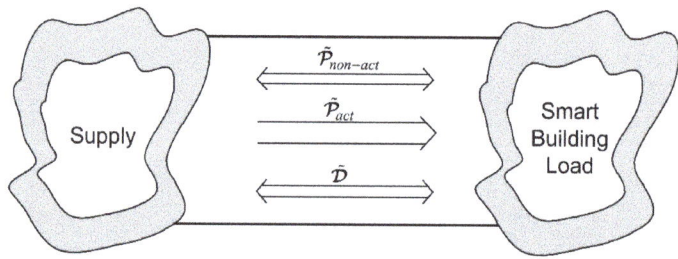

Figure 4. Flux of the generalized Poynting multivector components of electrical systems.

From Equation (10), the energy balance can be expressed as a multivector $\tilde{S}$ in $\{\mathcal{CG}_n\}$, generated by the geometric product "$\odot$" of the voltage and conjugate current geometric phasors [14] given by the following set

$$\tilde{S} = \tilde{U} \odot \tilde{I}^* = \tilde{U} \cdot \tilde{I}^* + \tilde{U} \wedge \tilde{I}^* \quad (14)$$

where $\tilde{U}$ is the voltage geometric phasor and $\tilde{I}^*$ is the conjugated current geometric phasor.

The electromagnetic object $\tilde{\mathcal{P}}$ can be separated into active and non-active components

$$\widetilde{\mathcal{P}} = \sum_p \widetilde{\mathcal{P}}_{p_{act}} + \sum_p \widetilde{\mathcal{P}}_{p_{non-act}} = \underbrace{\operatorname{Re}\left\{\sum_p \widetilde{\mathcal{P}}_p\right\}}_{\widetilde{\mathcal{P}}_{act}} + \underbrace{j\operatorname{Im}\left\{\sum_p \widetilde{\mathcal{P}}_p\right\}}_{\widetilde{\mathcal{P}}_{non-act}} \quad (15)$$

where the set $\widetilde{\mathcal{P}}_{act}$ transfers the harmonic active power and the set $\widetilde{\mathcal{P}}_{non-act}$ transport the scalar non-active power (classical reactive power). The only component of $\widetilde{\mathcal{P}}$ that transfer useful energy is $\widetilde{\mathcal{P}}_{act}$, Figure 4.

Hence, by virtue of (10), it is obtains

$$\iint_\varsigma \sum_p \vec{1}_z \cdot \widetilde{\mathcal{P}}_p \, d\varsigma = \iint_\varsigma \sum_p \vec{1}_z \cdot \widetilde{E}_p \odot \widetilde{H}_p^* \, d\varsigma = \sum_P U_p I_p e^{j\phi_p} \sigma_0 \quad (16)$$

The PM ($\widetilde{\mathcal{P}}$) is associated to the energy density at a point on the surface given in terms of the harmonic spatial geometric phasor of electric and magnetic fields at that point. Observe that the PM coincides with the classic Complex Poynting Vector only in pure sinusoidal case.

The real part of $\widetilde{\mathcal{P}}$ in (16), $P = \operatorname{Re}\left\{\iint_\varsigma \sum_p \vec{1}_z \cdot \widetilde{\mathcal{P}}_p \, d\varsigma\right\}$, permits a direct interpretation in terms of average energy flow, (i.e. active power P).

The imaginary part of $\widetilde{\mathcal{P}}$ in (16), $Q = \operatorname{Im}\left\{\iint_\varsigma \sum_p \vec{1}_z \cdot \widetilde{\mathcal{P}}_p \, d\varsigma\right\}$, is the scalar non-active power.

In general, it can be verified that PM $\widetilde{\mathcal{P}}$ considers direction and sense, not only of active power components $\widetilde{\mathcal{P}}_{p_{act}}$ but also of scalar non-active components $\widetilde{\mathcal{P}}_{p_{non-act}}$. Thus, the possible reverse sense of any harmonic $\widetilde{\mathcal{P}}_{p_{act}}$ is very important for a correct identification of harmonic source locations and for determining the responsibility of the utility electrical system and the associated load. Thus, if $\widetilde{\mathcal{P}}_{p_{act}} \geq 0$ ($P_p \geq 0$) the energy flow is unidirectional toward the load. On the other way, if $\widetilde{\mathcal{P}}_{p_{act}} \leq 0$ ($P_p \leq 0$) the flow is generated in the nonlinearity of the load.

A more detailed develop of the CPM ($\widetilde{\mathcal{D}}$) is

$$\widetilde{\mathcal{D}} = \sum_{p<q} \widetilde{E}_p \odot \widetilde{H}_q^* + \widetilde{E}_q \odot \widetilde{H}_p^* = \sum_{\substack{p<q \\ p,q \in N}} \widetilde{\mathcal{D}}_{pq} + \sum_{p \in N, q \in M} \widetilde{\mathcal{D}}_{pq}$$

$$= \underbrace{\operatorname{Re}\left\{\sum_{\substack{p<q \\ p,q \in N}} \widetilde{\mathcal{D}}_{pq} + \sum_{p \in N, q \in M} \widetilde{\mathcal{D}}_{pq}\right\}}_{\widetilde{\mathcal{D}}_{act}} + \underbrace{j\operatorname{Im}\left\{\sum_{\substack{p<q \\ p,q \in N}} \widetilde{\mathcal{D}}_{pq} + \sum_{p \in N, q \in M} \widetilde{\mathcal{D}}_{pq}\right\}}_{\widetilde{\mathcal{D}}_{non-act}} \quad (17)$$

Moreover, the flux of the Complementary Poynting Multivector is given by

$$\iint_\varsigma \sum_{p<q} \vec{1}_z \cdot \widetilde{\mathcal{D}}_{pq} \, d\varsigma = \widetilde{D} \quad (18)$$

where $\widetilde{D}$ can be associated to the polluting nature components, $\widetilde{\mathcal{D}}_{act}$ and $\widetilde{\mathcal{D}}_{non-act}$, that do not help in transferring useful energy and

$$\widetilde{D} = \sum_{\substack{p<q \\ linear}} \left\{\left(U_p I_q e^{j\varphi_q} - U_q I_p e^{j\varphi_p}\right) e^{j(\alpha_p - \alpha_q)}\right\} \sigma_{pq} + \sum_{\substack{p<q \\ non-linear}} U_p I_q e^{j(\alpha_p - \beta_q)} \sigma_{pq} \quad (19)$$

is the non-active power bivector called distortion power.

Through (9), (16) and (18), it can be observed that the energy flow that it is originated from the surface of the source in the electrical system equals the flow that enters in the load surface.

5.2. Electromagnetic Quality Index (EQI)

The aim of this paper is to propose a multivectorial index $\tilde{\zeta}$, based on electromagnetic geometric objects, capable to characterize the performance of the power transfer or the efficiency of the transmission equipment in single-phase circuits with multi-harmonic signals. This figure of merit can be an important piece of information for the interpretation of the electromagnetic energy flow in presence of disturbing loads. On the contrary, the classic THD index (Total Harmonic Distortion) for the measurement of the harmonic distortion level is not able to give any information about the disturbance nature.

To this end, a new index that integrally reflects the nature of the different energy quality situations is defined. This is expressed as

$$\tilde{\zeta} = \frac{\iint_\varsigma \vec{1}_z \tilde{\Pi} d\varsigma}{\text{Re}\left\{\iint_\varsigma \sum_{p \in N} \vec{1}_z \cdot \tilde{\mathcal{P}}_p \, d\varsigma\right\}} = 1 + j \frac{\text{Im}\left\{\iint_\varsigma \sum_{p \in N} \vec{1}_z \cdot \tilde{\mathcal{P}}_p \, d\varsigma\right\}}{\text{Re}\left\{\iint_\varsigma \sum_{p \in N} \vec{1}_z \cdot \tilde{\mathcal{P}}_p \, d\varsigma\right\}} + \frac{\iint_\varsigma \sum_{p \neq q} \vec{1}_z \cdot \tilde{\mathcal{D}}_{pq} \, d\varsigma}{\text{Re}\left\{\iint_\varsigma \sum_{p \in N} \vec{1}_z \cdot \tilde{\mathcal{P}}_p \, d\varsigma\right\}} \qquad (20)$$

or also,

$$\tilde{\zeta} = \frac{\text{Re}\left\{\iint_\varsigma \sum_{p=1} \vec{1}_z \cdot \tilde{\mathcal{P}}_1 \, d\varsigma\right\} + j\text{Im}\left\{\iint_\varsigma \sum_{p=1} \vec{1}_z \cdot \tilde{\mathcal{P}}_1 \, d\varsigma\right\}}{\text{Re}\left\{\iint_\varsigma \sum_p \vec{1}_z \cdot \tilde{\mathcal{P}}_p \, d\varsigma\right\}} +$$

$$+ \frac{\text{Re}\left\{\iint_\varsigma \sum_{p \neq 1} \vec{1}_z \cdot \tilde{\mathcal{P}}_p \, d\varsigma\right\} + j\text{Im}\left\{\iint_\varsigma \sum_{p \neq 1} \vec{1}_z \cdot \tilde{\mathcal{P}}_p \, d\varsigma\right\}}{\text{Re}\left\{\iint_\varsigma \sum_p \vec{1}_z \cdot \tilde{\mathcal{P}}_p \, d\varsigma\right\}} + \qquad (21)$$

$$+ \frac{\iint_\varsigma \sum_{p \neq q} \vec{1}_z \cdot \tilde{\mathcal{D}}_{pq} \, d\varsigma}{\text{Re}\left\{\iint_\varsigma \sum_p \vec{1}_z \cdot \tilde{\mathcal{P}}_p \, d\varsigma\right\}}$$

The power factor PF can be written as

$$PF = \frac{1}{\|\tilde{\zeta}\|} = \frac{\left\|\text{Re}\left\{\iint_s \sum_p \vec{1}_z \cdot \tilde{\mathcal{P}}_p \, ds\right\}\right\|}{\left\|\iint_s \vec{1}_z \cdot \tilde{\Pi} \, ds\right\|} \qquad (22)$$

It is noteworthy that the power factor is not an exhaustive index for energy quality. In fact, combining (20) and (22) the power factor could be brought to unity, but the electrical systems would still operate in multi-sinusoidal mode.

Equations (21), (22) and (23) show that the index $\tilde{\zeta}$ contains terms that keep direction and sense, allowing harmonic source detection. This property is very important to achieve an appropriated compromise between the energy quality index and the power factor. Moreover, the dominant harmonic source should be based on an evaluation of non-active scalar and bivector electromagnetic geometric object components of the Generalized Poynting Multivector. This subject is discussed in the next subsection.

5.3. Illustrative Comparison between Different Non-Active Electromagnetic Geometric Objects

It is seen from (12) and (17) that $\tilde{U}_1$ and $\tilde{I}_1$ are the sinusoidal voltage and current geometric phasor and $\tilde{U}_p$ and $\tilde{I}_q$ are the harmonic voltage and current geometric phasor when $p \neq 1, q \neq 1$ respectively. Obviously, the voltage geometric phasor $\tilde{U}_1$ produces the sinusoidal electric field spatial geometric phasor $\tilde{E}_1$ and the harmonic voltage $\tilde{U}_p$ generates the harmonic field $\tilde{E}_p$. Similarly the sinusoidal current $\tilde{I}_1$ geometric phasor produces the sinusoidal magnetic field spatial geometric phasor $\tilde{H}_1$ and the harmonic current $\tilde{I}_q$ generates the harmonic field $\tilde{H}_q$. The interaction among these fields produces the cited characteristic electromagnetic geometric objects $\tilde{\Pi}$, $\tilde{\mathcal{P}}$ and $\tilde{\mathcal{D}}$. The objects $\tilde{\mathcal{P}}$ and $\tilde{\mathcal{D}}$ are separated into an active and a non-active component as in Equations (15) and (17). In these equations, the different non-active scalar and bivector electromagnetic objects assume different expressions depending on the electrical system nature.

Starting from these considerations, a new philosophy is proposed for the detection of the dominant harmonic source that is based on the comparison of these non-active components to explain the power quality concept. For more detail, the following situations are considered:

- Sinusoidal case: $p = q = 1 \Rightarrow N = \{1\}$

$$\tilde{\Pi}^{\sin} = \tilde{\mathcal{P}}_1 = \underbrace{\text{Re}\{\tilde{\mathcal{P}}_1\}}_{\tilde{\mathcal{P}}_{1_{act}}} + j\underbrace{\text{Im}\{\tilde{\mathcal{P}}_1\}}_{\tilde{\mathcal{P}}_{1_{non-act}}} \qquad (23)$$

and

$$\tilde{\varsigma}^{\sin} = 1 + j \frac{\text{Im}\left\{\iint_\varsigma \sum_{p\in 1} \vec{1}_Z \cdot \tilde{\mathcal{P}}_1 \, d\varsigma\right\}}{\text{Re}\left\{\iint_\varsigma \sum_{p\in 1} \vec{1}_Z \cdot \tilde{\mathcal{P}}_1 \, d\varsigma\right\}} \qquad (24)$$

In this case, the quantity $\text{Im}\{\tilde{\mathcal{P}}_1\} = \tilde{\mathcal{P}}_{1_{non-act}}$ can be considered as minimum reference value to improve the energy quality, since it is the only non-active electromagnetic geometric object. It is the well established fundamental reactive power (Q) that can be reduced to zero by shunts capacitors

$$\left\|\text{Im}\left\{\iint_\varsigma \sum_{p\in 1} \vec{1}_Z \cdot \tilde{\mathcal{P}}_1 \, d\varsigma\right\}\right\| = 0 \qquad (25)$$

and consequently,

$$\left\|\tilde{\varsigma}^{\sin}\right\| = 1 \Rightarrow PF_1 = 1 \qquad (26)$$

where PF_1 is the fundamental power factor, also known as the displacement power factor

$$PF_1 = \cos(\varphi_1) \qquad (27)$$

The quantity $\text{Re}\{\tilde{\mathcal{P}}_1\} = \tilde{\mathcal{P}}_{1_{act}}$ transfers the fundamental active power (useful energy) and is associated to the instantaneous active power.

- Multi-sinusoidal linear case: $p \in N, q \in N$

If the electrical system is linear, the current requested by the loads and voltage supplied by the mains share the same harmonic order and the equation (8) yields the linear GPM $\widetilde{\Pi}^{lin}$

$$\widetilde{\Pi}^{lin} = \sum_{p=q} \widetilde{\mathcal{P}}_p + \sum_{p \neq q} \widetilde{\mathcal{D}}_{pq} =$$
$$= \underbrace{\sum_{p=q} \widetilde{\mathcal{P}}_{p_{act}}}_{\widetilde{\mathcal{P}}_{act}} + \underbrace{\widetilde{\mathcal{P}}_{1_{non-act}} + \sum_{p,q \in N > 1} \widetilde{\mathcal{P}}_{p_{non-act}}}_{j\widetilde{\mathcal{P}}_{non-act}} + \underbrace{\sum_{p,q \in N} \widetilde{\mathcal{D}}_{pq}}_{\widetilde{\mathcal{D}}^{lin}} \qquad (28)$$

The linear EQM is given by

$$\widetilde{\zeta}^{lin} = 1 + j \frac{\mathrm{Im}\left\{\iint_\varsigma \sum_{p \in N} \vec{1}_Z \cdot \widetilde{\mathcal{P}}_p \, d\varsigma \right\}}{\mathrm{Re}\left\{\iint_\varsigma \sum_{p \in N} \vec{1}_Z \cdot \widetilde{\mathcal{P}}_p \, d\varsigma \right\}} + \frac{\iint_\varsigma \sum_{p,q \in N} \vec{1}_Z \cdot \widetilde{\mathcal{D}}_{pq} \, d\varsigma}{\mathrm{Re}\left\{\iint_\varsigma \sum_{p \in N} \vec{1}_Z \cdot \widetilde{\mathcal{P}}_p \, d\varsigma \right\}} \qquad (29)$$

The quantities $\mathrm{Im}\left\{\iint_\varsigma \sum_{p \in N} \vec{1}_Z \cdot \widetilde{\mathcal{P}}_p \, d\varsigma \right\}$ and $\mathrm{Im}\left\{\iint_\varsigma \sum_{p,q \in N} \vec{1}_Z \cdot \widetilde{\mathcal{D}}_{pq} \, d\varsigma \right\}$ in (29) should be minimized by shunt capacitors or reduced to zero by shunt reactance one-ports. In these conditions

$$\widetilde{\zeta}^{lin} = 1 + \frac{\mathrm{Re}\iint_\varsigma \sum_{p,q \in N} \vec{1}_Z \cdot \widetilde{\mathcal{D}}_{pq} \, d\varsigma}{\mathrm{Re}\left\{\iint_\varsigma \sum_{p \in N} \vec{1}_Z \cdot \widetilde{\mathcal{P}}_p \, d\varsigma \right\}} \Rightarrow \left|\left|\widetilde{\zeta}^{lin}\right|\right| > 1 \Rightarrow PF < 1 \qquad (30)$$

From (30), it results that for linear electrical systems under multi-sinusoidal conditions, the criterion

$$\mathrm{Im}\left\{\iint_\varsigma \sum_{p \in N} \vec{1}_Z \cdot \widetilde{\mathcal{P}}_p \, d\varsigma \right\} = 0 \qquad (31)$$

$$\mathrm{Im}\left\{\iint_\varsigma \sum_{p,q \in N} \vec{1}_Z \cdot \widetilde{\mathcal{D}}_{pq} \, d\varsigma \right\} = 0 \qquad (32)$$

does not represent the conditions of highest power factor.

- Multi-sinusoidal non-linear case: $p \in N$, $q \in N \cup M$

In this case, the presence of the nonlinear loads causes some current components ($q \in M$) which harmonic orders are not present in the voltage supplied to the electrical system. It is well known that when a non-linear o time-varying electrical system is present, it injects disturbances even if the supply voltage is sinusoidal. In view of (8), the non-linear GPM can be written as follows,

$$\widetilde{\Pi}^{non-lin} = \sum_{p=q} \widetilde{\mathcal{P}}_p + \sum_{p \neq q} \widetilde{\mathcal{D}}_{pq} =$$
$$= \underbrace{\sum_{p=q} \widetilde{\mathcal{P}}_{p_{act}}}_{\widetilde{\mathcal{P}}_{act}} + \underbrace{j\widetilde{\mathcal{P}}_{1_{non-act}} + \sum_{p,q \in N > 1} \widetilde{\mathcal{P}}_{p_{non-act}}}_{j\widetilde{\mathcal{P}}_{non-act}} + \underbrace{\sum_{p,q \in N} \widetilde{\mathcal{D}}_{pq}}_{\widetilde{\mathcal{D}}^{lin}} + \underbrace{\sum_{\substack{p \in N, \\ q \in M}} \widetilde{\mathcal{D}}_{pq}}_{\widetilde{\mathcal{D}}^{non-lin}} \qquad (33)$$

The EQI is given by

$$\tilde{\zeta}^{non-lin} = 1 + j\frac{\text{Im}\left\{\iint_\varsigma \sum_{p\in N} \vec{1}_Z \cdot \tilde{\mathcal{P}}_p \, d\varsigma\right\}}{\text{Re}\left\{\iint_\varsigma \sum_{p\in N} \vec{1}_Z \cdot \tilde{\mathcal{P}}_p \, d\varsigma\right\}} + \frac{\iint_\varsigma \sum_{\substack{p\in N, q\in N \\ p\in N, q\in M}} \vec{1}_Z \cdot \tilde{\mathcal{D}}_{pq} \, d\varsigma}{\text{Re}\left\{\iint_\varsigma \sum_{p\in N} \vec{1}_Z \cdot \tilde{\mathcal{P}}_p \, d\varsigma\right\}} \quad (34)$$

If the equalities (31) and (32) are fulfilled, the Eq. (34) is now given by,

$$\tilde{\zeta}^{non-lin} = 1 + \frac{\text{Re}\iint_\varsigma \sum_{p,q\in N} \vec{1}_Z \cdot \tilde{\mathcal{D}}_{pq} \, d\varsigma + \iint_\varsigma \sum_{p\in N, q\in M} \vec{1}_Z \cdot \tilde{\mathcal{D}}_{pq} \, d\varsigma}{\text{Re}\left\{\iint_\varsigma \sum_{p\in N} \vec{1}_Z \cdot \tilde{\mathcal{P}}_p \, d\varsigma\right\}} \quad (35)$$

And

$$\left\|\tilde{\zeta}^{non-lin}\right\| > 1 \Rightarrow PF < 1 \quad (36)$$

Comparing the multivectors expressed in (26), (30) and (36), it can be observed that $\tilde{\zeta}^{non-lin}$ multivector contains all possible non-active components after passive compensation. Consequently, in the same operation conditions, it can be written that

$$\left\|\tilde{\zeta}^{\sin}\right\| \leq \left\|\tilde{\zeta}^{lin}\right\| \leq \left\|\tilde{\zeta}^{non-lin}\right\| \quad (37)$$

In sinusoidal operation (ideal situation), the three indexes are equal. In the presence of harmonic distortion, the differences among the values of the indexes depend of the supply and of the load characteristics. Then, the possible $\tilde{\zeta}$ multivectors and their magnitudes depend on the electrical system conditions. Each situation is strictly related to the distortion state of power system and therefore, to the harmonic presence and energy quality, thus, the higher harmonic interaction, the greater non-active energy flow. Observe that the multivectors $\tilde{\mathcal{P}}_{non-act}$ and $\tilde{\mathcal{D}}$ do not help in transferring useful energy and only are associated to the oscillations that produce both, energy lost and stored energy by the loads. Both quantities are related with the non-active power.

As a result of these considerations, an evaluation and comparison of both indexes, $\tilde{\zeta}$ and PF, in the same working conditions, shows that $\tilde{\zeta}$ give two pieces of extra information about the energy quality and the disturbing loads nature.

In conclusion, the suggested EQI possesses clear advantages from the viewpoint of non-active energy flow minimization. The main advantage is that it is decomposed into a complex-scalar and complex-bivectors with direction and sense. These components provide detailed information for a possible minimization of each electromagnetic object term by means of new devices, strategies, and algorithms. The accomplishment of such compensating methods and devices is a problem that warrants further research.

Although the proposed theory is limited to linear and non-linear distorted single-phase power systems, it is worth mentioning that this work does not have in any way the pretension to put an end to the topic, quite the contrary, in fact. Thus, the application of this methodology to polyphase systems deserves in-depth investigations in the near future. This proposed approximation has to face different unsolved problematic aspects. In particular, while the definition of apparent power and related components in balanced three-phase systems with sinusoidal waveforms is well established, the definitions of unbalance conditions are still in an open debate. Moreover, the study of the most general case, the three-phase systems with non-sinusoidal and unbalanced conditions, even needs to improve precedent theories that in many cases are mutually contradictory.

It is remarkable that the structure of geometric algebra gives a new insight to phase domain and constitutes a powerful tool for the treatment of periodic distorted signals. In this framework, power definition deals with the key concept of geometric phasors, i.e. algebraic time-averaged quantities that are far from being instantaneous ones. In this regard, depending on the considered application, this can be a major restriction. Thus, applications such as instantaneous active filters or fast response compensation devices, are out of the scope of this study, as well. For the same reason, electric signals disturbed with non-stationary events such as transients, sag, swells, etc. are not considered in this work.

6. Numerical Example

In order to validate the use and relevance of the Poynting Multivector suggested in this paper, sustainable and smart building loads supplied from a transmission line are analysed in the $\mathcal{CG}_n - \mathbb{R}^3$ framework. In a simplistic manner, the conductors are considered as rectangular and parallel superconductors meant to facilitate the propagation of energy from source to load, Figure 5. Units of physical quantities are the standard units of the MKSA system and thus are omitted.

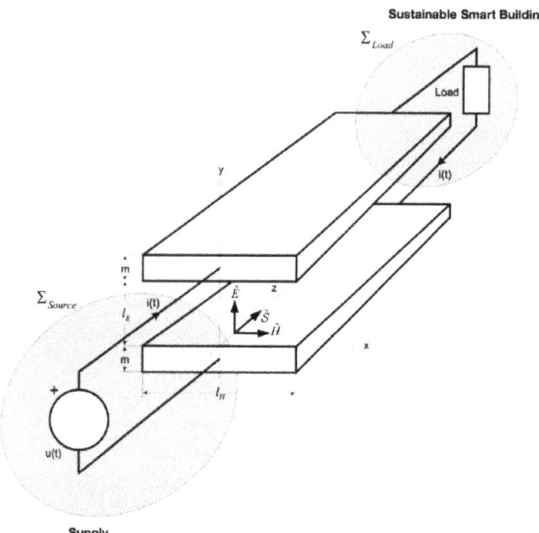

Figure 5. Industrial system: Sustainable smart building.

Two parallel plane conductors in linear media are considered. Both conductors, of thickness λ and width γ, are separated by a dielectric material of thickness ρ. We suppose that $\gamma \gg \lambda, \rho$. By ignoring eddy currents, line impedance, fringing effects, three cases are analysed:

6.1. Linear Load Supplied by A Sinusoidal Voltage Source

Voltage and a hypothetic resulting current are respectively

$$u(t) = \sqrt{2}(200 \sin \omega_1 t)$$

$$i(t) = \sqrt{2}[10 \sin(\omega_1 t - 36.87°)]$$

Then, from [13], the spatial geometric phasors of the electric $\tilde{E}$ and magnetic $\tilde{H}^*$ fields can be expressed as

$$\tilde{E} = \frac{1}{d}\left(200e^{j0}\sigma_1\right)\vec{1}_X ; \tilde{H}^* = \frac{1}{h}\left(10e^{j36.87}\sigma_1\right)\vec{1}_Y$$

and the electromagnetic geometric objects and associated power are

$$\iint_\varsigma \sum_{p=1} \vec{1}_Z \cdot \tilde{\mathcal{P}} \, d\varsigma = \left(\underbrace{1600}_{P_1} + j\underbrace{1200}_{Q_1} \right) \sigma_0$$

$$\iint_\varsigma \sum_{p \neq q} \vec{1}_Z \cdot \tilde{\mathcal{D}} \, d\varsigma = 0; \tilde{D} = 0$$

$$\tilde{\xi}^{sin} = 1 + j\frac{1200}{1600}\sigma_0 \Rightarrow \left\lVert \tilde{\xi}^{sin} \right\rVert = 1.25; PF = 0.8$$

6.2. Linear Load Supplied by A Non-Sinusoidal Sinusoidal Voltage Source

Voltage and hypothetic resulting current are

$$u(t) = \sqrt{2}(200 \sin \omega_1 t + 50 \sin \omega_2 t)$$

$$i(t) = \sqrt{2}[10 \sin(\omega_1 t - 36.87°) + 5 \sin(\omega_2 t + 53.13°)]$$

and the spatial geometric phasors of the electric $\tilde{E}$ and magnetic $\tilde{H}^*$ fields can be expressed as

$$\tilde{E} = \frac{1}{d}\left(200e^{j0}\sigma_1 + 50e^{j0}\sigma_2 \right) \vec{1}_X$$

$$\tilde{H}^* = \frac{1}{h}\left(10e^{j36.87}\sigma_1 + 5e^{-j53.13}\sigma_2 \right) \vec{1}_Y$$

$$\iint_\varsigma \sum_p \vec{1}_Z \cdot \tilde{\mathcal{P}} \, d\varsigma = \left(\underbrace{1750}_{P} + j\underbrace{1000}_{Q} \right) \sigma_0$$

$$\iint_\varsigma \sum_{p \neq q} \vec{1}_Z \cdot \tilde{\mathcal{D}} \, d\varsigma = \tilde{D} = \underbrace{[200 - j1100]\sigma_{12}}_{\tilde{D}_{12}}$$

$$\tilde{\xi}^{non-lin} = 1 + j\frac{1000}{1750}\sigma_0 + \frac{[200 - j1100]\sigma_{12}}{1750\sigma_0}$$

$$\left\lVert \tilde{\xi}^{non-lin} \right\rVert = 1.32$$

$$PF = 0.76$$

6.3. Non-Linear Load Supplied by A Non-Sinusoidal Sinusoidal Voltage Source

Voltage and hypothetic resulting current are

$$u(t) = \sqrt{2}(200 \sin \omega_1 t + 50 \sin \omega_2 t)$$

$$i(t) = \sqrt{2}[10 \sin(\omega_1 t - 36.87°) + 5 \sin(\omega_2 t + 53.13°) + 5 \sin(\omega_3 t + 45°)]$$

and the spatial geometric phasors of the electric $\tilde{E}$ and magnetic $\tilde{H}^*$ fields can be expressed as

$$\tilde{E} = \frac{1}{d}\left(200e^{j0}\sigma_1 + 50e^{j0}\sigma_2 \right) \vec{1}_X$$

$$\tilde{E} = \frac{1}{d}\left(200e^{j0}\sigma_1 + 50e^{j0}\sigma_2 \right) \vec{1}_X$$

$$\iint_\varsigma \sum_p \vec{1}_z \cdot \tilde{\mathcal{P}} \, d\varsigma = [(1600+1200)+j(150-200)]\sigma_0 = \Big(\underbrace{1750}_{P}+j\underbrace{1000}_{Q}\Big)\sigma_0$$

$$\iint_\varsigma \sum_{p\neq q} \vec{1}_z \cdot \tilde{\mathcal{D}} \, d\varsigma = \tilde{D} = \underbrace{[200-j1100]}_{\tilde{D}_{12}}\sigma_{12} + \underbrace{\left[125\sqrt{2}-j125\sqrt{2}\right]}_{\tilde{D}_{23}}\sigma_{13} + \underbrace{\left[500\sqrt{2}-j500\sqrt{2}\right]}_{\tilde{D}_{13}}\sigma_{23}$$

$$\tilde{\xi}^{non-lin} = 1 + j\frac{1000}{1750}\sigma_0 + \frac{[200-j1100]\sigma_{12} + \left[125\sqrt{2}-j125\sqrt{2}\right]\sigma_{13} + \left[500\sqrt{2}-j500\sqrt{2}\right]\sigma_{23}}{1750\sigma_0}$$

$$\left\|\tilde{\xi}^{non-lin}\right\| = 1.44$$

$$PF = 0.69$$

As in (37), it should be noticed in this example that

$$\left\{\left\|\tilde{\xi}^{\sin}\right\| = 1.25\right\} < \left\{\left\|\tilde{\xi}^{lin}\right\| = 1.32\right\} < \left\{\left\|\tilde{\xi}^{non-lin}\right\| = 1.44\right\}$$

Comparing the multivectors $\tilde{\xi}$ of the above three cases, it can be observed that they contain all possible non-active components. In this way, the possible reverse sense of any harmonic of active, reactive and distortion powers is very important for a correct identification of harmonic source locations, and for determining the responsibility of the utility side (generator) and the consumer side (load) [20,21]. Thus, the suggested index $\tilde{\xi}$ possesses clear advantages from the viewpoint of non-active power flow minimization.

7. Conclusions

The analysis of sustainable smart buildings is key to new future buildings, new complex architectures, and its usefulness extends to smart cities. Analyses of quality of the construction typically focus on applying methodologies that evaluate quality objectives at environmental, construction and building levels. Research has shown that a multivector quality index can be useful for detection of harmonic sources of new sustainable smart buildings.

Along this line this paper presents a reformulation of Poynting Vector in terms of the Geometric Algebra framework when inefficiencies caused by harmonics are considered. In this environment, each geometric object represents a different kind of energy flux. Thus, the distinct and well-known power terms in electric power systems, i.e. Active, Reactive, Distortion and Apparent powers acquire a new interpretation and dimension by means of multivectors. This novel approach is applicable to smart architectural single-phase power systems with linear/non-linear loads under sinusoidal or non-sinusoidal conditions. By this means, the behaviour of the energy flux is summarized by the proposed electromagnetic energy index EQI. In addition, the meaning of this original index is deeply analysed and discussed, concluding that its capability to deal with geometric properties, namely magnitude, direction and sense, makes it perfectly appropriate for detection and minimization of harmonic sources.

Furthermore, it is an excellent tool to interpret the cited power flow distribution. Thus, the introduction of this figure of merit supposes not only a generalization of the traditional power factor in electric power networks but also the key to the electromagnetic energy transfer between mains and loads. The global efficiency of a system is truth evaluated by this index because it includes all possible interactions.

It is remarkable that traditional Poynting Vector is a particular case of the proposed generalized Poynting Multivector when no cross interactions are present between different harmonics. This correlation is now precisely understood as the outer product between geometric phasors of different energy levels. Thus, this fact is crucial for future research proposals on the design of passive and/or active architectures for selective compensation, optimization algorithms and measurement techniques for harmonic pollution monitoring.

The electromagnetic nature of the power components could be the key that opens a broad range of new interesting research lines in electrical engineering and all energy efficiency related matters.

Author Contributions: All authors contributed equally to this work. All authors wrote, reviewed and commented on the manuscript. All authors have read and approved the final manuscript.

Conflicts of Interest: The authors declare no conflict of interest.

Glossary of Symbols

$\mathbb{R}$	real numbers
$\mathcal{C}$	complex vector space
$\mathcal{G}_n$	Geometric Algebra in n-dimensional real space
$\mathcal{CG}_n$	Complex Geometric Algebra
$\mathcal{CG}_n^t - \mathbb{R}^3$	time generalized geometric euclidean space
$\mathcal{CG}_n - \mathbb{R}^3$	frequency generalized geometric euclidean space
$\vec{1}_X, \vec{1}_Y, \vec{1}_Z$	Euclidean canonical basis
$\sigma_{1...k}$	canonical basis of $\mathcal{G}_n$
σ_p	basis vector of $\mathcal{G}_n$
$\sigma_p \sigma_q = \sigma_{pq}$	basis bivectors of $\mathcal{G}_n$
$\sigma_1 \sigma_2 \sigma_3$	trivector or pseudoscalar of $\mathcal{G}_n$
λ, μ	scalars or 0-grade geometric object
a, b	vectors or 1-grade geometric object
B	bivector or 2-grade geometric object
J	pseudoscalar or n-grade geometric object
$\widetilde{M}$	generic multivector
$\widetilde{\Pi}$	generalized Poynting multivector (GPM)
$\widetilde{\mathcal{P}}$	Poynting multivector (PM)
$\widetilde{\mathcal{D}}$	Complementary Poynting multivector (CPM)
$\widetilde{E}$	electric field geometric phasor
$\widetilde{H}$	magnetic field geometric phasor
$\odot$	generalized geometric product
$\cdot$	inner product
$\wedge$	outer product
j	imaginary unit
$*$	conjugated operation
$\dagger$	reverse operation
$\sim$	multivector characterization
U_p	p-th harmonic voltage rms value
I_q	q-th harmonic current rms value
P	active power or real part of 0-grade power multivector
Q	reactive power or imaginary part of 0-grade power multivector
D	distortion power or 2-grade power
$\widetilde{S}$	power multivector
$\|\widetilde{S}\|, S$	apparent power multivector
ω_p	p-th harmonic frequency
α_p	phase angle of p-th voltage geometric phasor
β_q	phase angle of q-th current geometric phasor
ϕ_q	q-th impedance phase angle
$\widetilde{\xi}$	electromagnetic quality index multivector (EQI)
PF	power factor

References

1. Ibrahim, M.; El-Zaart, A.; Adams, C. Smart Sustainable Cities roadmap: Readiness for transformation towards urban sustainability. *Sustain. Cities Soc.* **2018**, *37*, 530–540. [CrossRef]
2. Höjer, M.; Wangel, J. Smart sustainable cities: Definition and challenges. In *ICT Innovations for Sustainability*; Hilty, L., Aebischer, B., Eds.; Springer: Cham, Switzerland, 2015; pp. 333–349.
3. Pervez, H.S.; Nursyarizal, B.M.N.; Perumal, N.; Irraivan, E.; Taib, I. A review on optimized control systems for building energy and comfort management of smart sustainable buildings. *Renew. Sustain. Energy Rev.* **2014**, *34*, 409–429. [CrossRef]
4. Nardi, I.; Lucchi, E.; Rubeis, T.; Ambrosini, D. Quantification of heat energy losses through the building envelope: A state-of-the-art analysis with critical and comprehensive review on infrared thermography. *Build. Environ.* **2018**, *146*, 190–205. [CrossRef]
5. Karmellos, M.; Kiprakis, A.; Mavrotas, G. A multi-objective approach for optimal prioritization of energy efficiency measures in buildings: Model, software and case studies. *Appl. Energy.* **2015**, *139*, 131–150. [CrossRef]
6. Poynting, J.H. On the transfer of energy in the electromagnetic field. *Philos. Trans. R. Soc.* **1884**, *175*, 343–361. [CrossRef]
7. Steinmetz, C.P. *Theory and Calculation of Alternating Current Phenomena*; McGraw Publishing Company: New York, NY, USA, 1900.
8. Emanuel, A.E. Powers in nonsinusoidal situations a review of definitions and physical meaning. *IEEE Trans. Power Deliv.* **1990**, *5*, 1377–1389. [CrossRef]
9. Emanuel, A.E. Poynting vector and the physical meaning of nonactive powers. *IEEE Trans. Instrum. Meas.* **2005**, *54*, 1457–1462. [CrossRef]
10. Lundin, U.; Bolund, B.; Leijon, M. Poynting vector analysis of synchronous generators using field simulations. *IEEE Trans. Magn.* **2007**, *43*, 3601–3606. [CrossRef]
11. Faria, J.A.B. The role of Poynting's vector in polyphase power calculations. *Eur. Trans. Electr. Power* **2009**, *19*, 683–688. [CrossRef]
12. De León, F.; Cohen, J. AC power theory from Poynting Theorem: Accurate identification of instantaneous power components in nonlinear-switched circuits. *IEEE Trans. Power Deliv.* **2010**, *25*, 2104–2112. [CrossRef]
13. Cheng, P.; Yang, F.; Luo, H.; Guo, H.; Ran, W.; Yang, Y.; Ullah, I. A method to calculate the reactive power of iced transmission line based on Poynting vector and FDFD. *Int. J. Appl. Electromagn. Mech.* **2016**, *50*, 417–433. [CrossRef]
14. Emanuel, A.E. About the Rejection of Poynting Vector in Power Systems Analysis. *Electr. Power Qual. Util. J.* **2007**, *13*, 43–49.
15. Emanuel, A.E. *Power Definitions and the Physical Mechanism of Power Flow*; Wiley: Hoboken, NJ, USA, 2010.
16. Todeschini, G.; Emanuel, A.E.; Ferrero, A.; Morando, A.P. A Poynting vector approach to the study of the Steinmetz compensator. *IEEE Trans. Power Deliv.* **2007**, *22*, 1830–1833. [CrossRef]
17. Ferrero, A.; Leva, S.; Morando, A.P. An approach to the non-active power concept in terms of the poynting-park vector. *Eur. Trans. Electr. Power* **2001**, *11*, 291–299. [CrossRef]
18. Czarnecki, L.S. Considerations on the Concept of Poynting Vector Contribution to Power Theory Development. In Proceedings of the 6th International Workshop on Power Definitions and Measurements Under Nonsinusoidal Conditions, Milano, Italy, 13–15 October 2003.
19. Czarnecki, L.S. Could Power Properties of Three-Phase Systems Be Described in Terms of the Poynting Vector? *IEEE Trans. Power Deliv.* **2006**, *21*, 339–344. [CrossRef]
20. Balci, M.; Hocaoglu, M.; Power, S.A. Transition from Poynting vector to instantaneous power. In Proceedings of the 12th International Conference on Harmonics and Quality of Power, Cascais, Portugal, 1–5 October 2006.
21. Stahlhut, J.W.; Browne, T.J.; Heydt, G.T. The Assessment of the Measurement of the Poynting Vector for Power System Instrumentation. In Proceedings of the 39th North American Power Symposium, Las Cruces, NM, USA, 30 September–2 October 2007; pp. 237–241.
22. Castilla, M.; Bravo, J.C.; Ordonez, M.; Montano, J.C. An approach to the multivectorial apparent power in terms of a generalized poynting multivector. *Prog. Electromagn. Res. B* **2009**, *15*, 401–422. [CrossRef]
23. Clifford, W.K. Applications of Grassmann's Extensive Algebra. *Am. J. Math.* **1878**, *1*, 350. [CrossRef]

24. Hestenes, D. A unified language for mathematics and physics. *their Appl. Math. Phys.* **1986**. [CrossRef]
25. Chappell, J.M.; Drake, S.P.; Seidel, C.L.; Gunn, L.J.; Iqbal, A.; Allison, A.; Abbott, D. Geometric algebra for electrical and electronic engineers. *Proc. IEEE* **2014**. [CrossRef]
26. Doran, C.; Lasenby, A. *Geometric Algebra for Physicists*; Cambridge University Press: Cambridge, UK, 2009.
27. Castilla, M.; Bravo, J.C.; Ordóñez, M.; Montaño, J.C. Clifford theory: A geometrical interpretation of multivectorial apparent power. *IEEE Trans. Circuits Syst. I Regul. Pap.* **2008**, *55*, 3358–3367. [CrossRef]
28. Castilla, M.; Bravo, J.C.; Ordonez, M.; Montaño, J.C.; Lopez, A.; Borras, D.; Gutierrez, J. The geometric algebra as a power theory analysis tool. In Proceedings of the 2008 International School on Nonsinusoidal Currents and Compensation, Lagow, Poland, 10–13 June 2008.
29. Castilla, M.; Bravo, J.C.; Ordóñez, M. Geometric algebra: A multivectorial proof of Tellegen's theorem in multiterminal networks. *IET Circuits, Devices Syst.* **2008**, *2*, 383–390. [CrossRef]
30. Bravo, J.C.; Castilla, M.V. Energy Conservation Law in Industrial Architecture: An Approach through Geometric Algebra. *Symmetry* **2016**, *8*, 92. [CrossRef]
31. Castro-Núñez, M.; Londoño-Monsalve, D.; Castro-Puche, R.M. the conservative power quantity based on the flow of energy. *J. Eng.* **2016**, *2016*, 269–276. [CrossRef]
32. Petroianu, A.I. A geometric algebra reformulation and interpretation of Steinmetz's symbolic method and his power expression in alternating current electrical circuits. *Electr. Eng.* **2015**, *97*, 175–180. [CrossRef]
33. Menti, A.; Zacharias, T.; Milias-Argitis, J. Geometric Algebra: A Powerful Tool for Representing Power Under Nonsinusoidal Conditions. *IEEE Trans. Circuits Syst. I Regul. Pap.* **2007**, *54*, 601–609. [CrossRef]
34. Castro-Nunez, M.; Castro-Puche, R. Advantages of Geometric Algebra Over Complex Numbers in the Analysis of Networks With Nonsinusoidal Sources and Linear Loads. *IEEE Trans. Circuits Syst. I Regul. Pap.* **2012**, *59*, 2056–2064. [CrossRef]
35. Herrera, R.S.S.; Salmeron, P.; Vazquez, J.R.R.; Litran, S.P.P.; Perez, A. Generalized instantaneous reactive power theory in poly-phase power systems. In Proceedings of the 13th European Conference on Power Electronics and Applications, Barcelona, Spain, 8–10 September 2009; pp. 1–10.

© 2018 by the authors. Licensee MDPI, Basel, Switzerland. This article is an open access article distributed under the terms and conditions of the Creative Commons Attribution (CC BY) license (http://creativecommons.org/licenses/by/4.0/).

Article

Denoising of Magnetocardiography Based on Improved Variational Mode Decomposition and Interval Thresholding Method

Yanping Liao, Congcong He * and Qiang Guo

Department of Information and Communication Engineering, Harbin Engineering University, Harbin 150001, China; liaoyanping@hrbeu.edu.cn (Y.L.); guoqiang@hrbeu.edu.cn (Q.G.)
* Correspondence: he_congcongs@126.com; Tel.: +86-188-4510-3530

Received: 23 May 2018; Accepted: 5 July 2018; Published: 9 July 2018

Abstract: Recently, magnetocardiography (MCG) has attracted increasing attention as a non-invasive and non-contact technique for detecting electrocardioelectric functions. However, the severe background noise makes it difficult to extract information. Variational Mode Decomposition (VMD), which is an entirely non-recursive model, is used to decompose the non-stationary signal into the intrinsic mode functions (IMFs). Traditional VMD algorithms cannot control the bandwidth of each IMF, whose quadratic penalty lacks adaptivity. As a result, baseline drift noise is still present or medical information is lost. In this paper, to overcome the unadaptable quadratic penalty problem, an improved VMD model via correlation coefficient and new update formulas are proposed to decompose MCG signals. To improve the denoising precision, this algorithm is combined with the interval threshold algorithm. First, the correlation coefficient is calculated, to determine quadratic penalty, in order to extract the first IMF made up of baseline drift. Then, the new update formulas derived from the variance that describes the noise level are used, to perform decomposition on the rest signal. Finally, the Interval thresholding algorithm is performed on each IMF. Theoretical analysis and experimental results show that this algorithm can effectively improve the output signal-to-noise ratio and has superior performance.

Keywords: magnetocardiography; quadratic penalty; variational mode decomposition; correlation coefficient; interval thresholding method

1. Introduction

In recent years, the research on signal processing, modeling, imaging theory, and methods related to bio-electromagnetism has become a hot topic. With the efforts of many experts and scholars, this field already has high-level research results. The magnetocardiography [1] signal plays an increasingly important role in heart disease diagnosis, which is detected with Superconducting Quantum Interference Devices (SQUID) [2] and has considerable advantages over electrocardiography (ECG) [3]. As the detecting instrument of the magnetocardiography signal, the SQUID operates from low to high temperature, and changes the number of channels from the original single channel into multiple channels. Magnetocardiography signals transmitted to the human chest surface are incredibly helpful toward both cardiac model reconstruction and clinical application [4,5]. The relationship between heart function and heart disease is studied by researching the characteristics of magnetic field strength changes at different locations; this type of study can be called interdisciplinary basic research. Generally, such measurements are conducted in order to detect small magnetic field signals in the presence of large background noise [6,7]. Removing background noise and recovering useful signals are chief objectives. For periodic signals, adequate suppression of uncorrelated noise may often be achieved by the signal averaging method [8]. If the signal and noise have separate bandwidths, one can use conventional frequency domain filtering [9] techniques. Adaptive filtering techniques [10] measure the noise level of the measurement signal channels by measuring the noise channels. However, these simple preprocessing methods have limited effectiveness. The wavelet transform method [11] for signal denoising is based on the use of a set of predefined basis functions, in order to decompose the measured signals and remove components corresponding to noise. The main disadvantage of this method is that the selection of wavelet basis seriously affects the denoising results. Empirical Mode Decomposition (EMD) [12,13] is one of the decomposition methods of signal denoising, and is widely used to decompose a signal into different modes recursively. This method is, however, prone to mode mixing, and limited by sensitivity to noise and sampling [14]. The mode mixing is significantly reduced by a modified noise-assisted data analysis method known as the Ensemble Empirical Mode Decomposition (EEMD) method [14,15]. The denoising principle of magnetocardiography (MCG) signals by EEMD based methods was reported in [16,17]. However, the decomposition results were unsatisfactory because of the low signal-to-noise ratio. In addition, the decomposition results of EMD and EEMD heavily depend on the extremum seeking algorithm and the ending criterion. A lack of mathematical approach and predefined filter boundaries reduce the accuracies of such detections [18]. Lately, based on the definition of intrinsic mode functions (IMF), a new adaptive decomposition method called Variational Mode Decomposition (VMD) [19] has been proposed. Supporting documents [20] proposed using the VMD method to denoise ECG signals. However, the research results showed that the decomposition results lack adaptability. The results of studies [21,22] showed that baseline drift noise was not filtered out by the VMD method. In practice, it is not always possible to have the first IMF to be a noise-only IMF.

In order to overcome the problems above, we propose an improved VMD method that determines the bandwidth of modes adaptively via the optimized quadratic penalty. The proposed correlation coefficient, between the IMF obtained by VMD [19,23] and the baseline drift model, is calculated repeatedly until the criterion is satisfied and the baseline drift noise is extracted. The new IMFs are then obtained by using new proposed update formulas that can be deduced by the relationship between the penalty factor and noise. The interval threshold method is used for the subsequent processing of each component, which removes noise components.

The rest of this paper is organized as follows: Section 2 introduces the data model required for the VMD algorithm. In Section 3, a new VMD scheme based on the correlation coefficient and new updated formulas is proposed. The application for denoising methods of MCG is shown in Section 4. Conclusions are given in Section 5.

2. Data Model

In the expression of the traditional EMD and EEMD methods, the IMF is defined as a function where the difference between the number of zeros and poles does not exceed one [24]. In recent studies, the definition of the modality is changed to amplitude-modulated-frequency-modulated (AM-FM) signal, defined as follows:

$$u_k(t) = A_k(t)\cos(\phi_k(t)) \tag{1}$$

In the above equation, the phase $\phi_k(t)$ is a non-decreasing function, whose first derivative is $\phi'_k(t) > 0$, where the envelope $A_k(t)$ is non-negative; both the envelope $A_k(t)$ and the instantaneous frequency $\omega_k(t)$ vary much slower than the phase $\phi_k(t)$ [25,26].

The Hilbert transform [27] is the convolution of a real function and the corresponding impulse response of $h(t) = 1/\pi t$ in time domain. It is an all-pass filter, characterized by the transfer function $H(\omega) = -j\text{sgn}(\omega) = -j\omega/|\omega|$ in frequency domain. The Hilbert transform of a purely real IMF $u_k(t)$ can be expressed as $\widetilde{u}_k(t)$, and the complex-valued analytic signal is now defined as:

$$\begin{aligned} u_{k,A}(t) = u_k(t) + j\widetilde{u}_k(t) &= A_k(t)[\cos(\phi(t)) - j\sin(\phi(t))] \\ &= A_k(t)e^{-j\phi(t)} \end{aligned} \tag{2}$$

where $\phi(t)$ is the phase, while the amplitude is governed by the real envelope. The expression $A_k(t)$. $\omega(t) = d\phi(t)/dt$ is the instantaneous frequency. The amplitude A_k for kth IMF signal changes slowly enough.

Research has shown, on a sufficiently long interval, that the mode can be considered to be a purely harmonic signal. In other words, the newer definition of signal components is slightly more restrictive than the original one, and the VMD mode is the particular case of the EMD mode.

3. Proposed New VMD Scheme

VMD as a new decomposition method, is a process to solve variational problems based on classic Wiener filtering and Hilbert transformation. We can use the VMD method to decompose a multi-component signal into several band-limited modes non-recursively, which are redefined as IMFs. However, the VMD algorithm cannot extract baseline drift noise when decomposing MCG signals. As such, a new VMD method framework is proposed in this paper.

3.1. Eliminate Baseline Drift Noise Using Proposed Formulas

To overcome the unadaptable quadratic penalty problem, we propose an improved VMD method with correlation coefficient and new update formulas. First, we need to extract the expected baseline drift noise that will be included in the first mode. The steps are given as follows:

1. Compute the associated analytic signal of each mode u_k by means of the Hilbert transform, that is:

$$u_{k,A}(t) = \left(\delta(t) + \frac{j}{\pi t}\right) * u_k(t) \qquad (3)$$

2. Mix each mode with an exponential adjustment to the respective estimated center frequency in order to shift the mode spectrum to "baseband".

$$\hat{u}_{k,A}(t) = \left[\left(\delta(t) + \frac{j}{\pi t}\right) * u_k(t)\right] e^{-j\omega_k t} \qquad (4)$$

where ω_k is the center frequency of the kth IMF $u_k(t)$.

3. Estimate the bandwidth through the squared L^2-norm of the gradient. The expression of the constrained variational problem is as follows:

$$\begin{cases} \min_{\{u_k\},\{\omega_k\}} \left\{ \sum_k \left\| \partial_t \left[\left(\delta(t) + \frac{j}{\pi t}\right) * u_k(t) \right] e^{-j\omega_k t} \right\|_2^2 \right\} \\ \text{s.t.} \quad \sum_{k=1}^{K} u_k(t) = f(t) \end{cases} \qquad (5)$$

where $\{u_k\} = \{u_1, \ldots, u_K\}$ and $\{\omega_k\} = \{\omega_1, \ldots, \omega_K\}$ are shorthand notations for the set of all modes and their center frequencies. In order to render the problem unconstrained, a quadratic penalty term α and Lagrangian multiplier λ are brought in. The quadratic penalty can encourage reconstruction fidelity, typically in the presence of additive Gaussian noise. The Lagrange equation can enforce constraints strictly. Therefore, we introduce the augmented Lagrange equation $\mathcal{L}$ as follow [28]:

$$\mathcal{L}(\{u_k\},\{\omega_k\},\lambda(t)) = \alpha \sum_{k=1}^{K} \left\| \partial_t \left[\left(\delta(t) + \frac{j}{\pi t}\right) * u_k(t) \right] e^{-j\omega_k t} \right\|_2^2 + \left\| f(t) - \sum_{k=1}^{K} u_k(t) \right\|_2^2 + \left\langle \lambda(t), f(t) - \sum_{k=1}^{K} u_k(t) \right\rangle \qquad (6)$$

Alternate direction method of multipliers (ADMM) is brought to solve the original minimization problem [29–31]. To update the mode u_k, we can get the equivalent minimization problem as the following:

$$u_k^{n+1} = \underset{u_k \in X}{\operatorname{argmin}} \left\{ \alpha \sum_{k=1}^{K} \left\| \partial_t \left[\left(\delta(t) + \frac{j}{\pi t} \right) * u_k(t) \right] e^{-j\omega_k t} \right\|_2^2 + \left\| f(t) - \sum_{i=1}^{K} u_i(t) - \frac{\lambda(t)}{2} \right\|_2^2 \right\} \quad (7)$$

where n is the number of iterations. Now, making use of the Parseval/Plancherel Fourier isometry under the L^2 norm, this problem can be solved in spectral domain. Then, performing a change of variables $\omega \leftarrow \omega - \omega_k$ in the first term, we can get the following expression:

$$\hat{u}_k^{n+1} = \underset{\hat{u}_k, u_k \in X}{\operatorname{argmin}} \left\{ \alpha \| j(\omega - \omega_k)[(1 + \operatorname{sgn}(\omega))\hat{u}_k(\omega)] \|_2^2 + \left\| \hat{f}(\omega) - \sum_i \hat{u}_i(\omega) + \frac{\hat{\lambda}(\omega)}{2} \right\|_2^2 \right\} \quad (8)$$

Exploiting the Hermitian symmetry of the real signals, we can write both terms as half-space integrals, then making the negative frequencies of the first variation disappeared as follows:

$$\hat{u}_k^{n+1}(\omega) = \frac{\hat{f}(\omega) - \sum_{i<k} \hat{u}_i^{n+1}(\omega) - \sum_{i>k} \hat{u}_i^n(\omega) + \frac{\hat{\lambda}^n(\omega)}{2}}{1 + 2\alpha(\omega - \omega_k^n)^2} \quad (9)$$

This is clearly identified as Wiener filtering of the current residual, with signal prior $1/(\omega - \omega_k^n)^2$; the time domain mode is obtained as the real part of the inverse Fourier transform of this filtered analytic signal. In order to obtain each component, the center frequency, corresponding to each component, needs to be solved. The center frequency appears in the bandwidth prior, but not in the reconstruction fidelity term. The relevant problem thus reads:

$$\omega_k^{n+1} = \underset{\omega_k}{\operatorname{argmin}} \left\{ \left\| \partial_t \left[\left(\delta(t) + \frac{j}{\pi t} \right) * u_k(t) \right] e^{-j\omega_k t} \right\|_2^2 \right\} \quad (10)$$

As before, the optimization can be taken place in Fourier domain, and we end up optimizing:

$$\omega_k^{n+1} = \frac{\int_0^\infty \omega |\hat{u}_k(\omega)|^2 d\omega}{\int_0^\infty |\hat{u}_k(\omega)|^2 d\omega} \quad (11)$$

In general, the baseline drift frequency is lower than the low frequency component of the MCG signal. The center frequency of the first IMF is approximately zero, and we need to reduce the bandwidth of the first IMF until the signal and baseline drift are separated. We need to know that the larger the penalty factor, the narrower the mode bandwidth. After completing the above iterative process, we can get the final $\hat{u}_k(\omega)$ and ω_k. In order to extract low-frequency baseline drift noise, we need to follow the above process to obtain the first IMF:

$$u_1(t) = e^{j\omega_1 t} \left(\frac{1}{2\pi} \int_{-\infty}^{\infty} \hat{u}_1(\omega) e^{j\omega t} d\omega \right) \quad (12)$$

To understand the relationship between the first IMF and baseline drift noise, we propose the correlation coefficient to estimate the relationship. Assuming that the baseline drift noise model is $u_1{'}(t)$ and the correlation coefficient ρ' are calculated by Equation (13):

$$\rho' = \frac{\sum_i (u_{1i}{'} - \overline{u_1{'}})(u_{1i} - \overline{u_1})}{\sqrt{\sum_i (u_{1i}{'} - \overline{u_1{'}})^2 \sum_i (u_{1i} - \overline{u_1})^2}} \quad (13)$$

In the above formula, the variable is simplified for convenience. Where i represents the length of the data, $\overline{u_1}'$ and $\overline{u_1}$ represent the mean of the baseline drift noise model and the first IMF, respectively. We believe that signal and baseline drift noise can be separated when the correlation coefficient reaches a certain threshold ρ. We increase α according to the proposed formulas to repeat the above process until satisfying $\rho' > \rho$. This satisfies the following formulas:

$$\begin{cases} imf_1 = u_1(t) & if \rho' > \rho \\ \alpha^{m+1} = \alpha^m + c_1, \hat{u}_1^{n+1}(\omega) = \dfrac{\hat{f}(\omega) - \sum_{i<1} \hat{u}_i^{n+1}(\omega) - \sum_{i>1} \hat{u}_i^n(\omega) + \frac{\lambda^n(\omega)}{2}}{1 + 2\alpha^{m+1}(\omega - \omega_1^n)^2} & if\ \rho' < \rho \end{cases} \quad (14)$$

where c_1 is a constant and m is the number of loop decomposition. After each updating of modes and center frequencies, the Lagrange multiplier $\hat{\lambda}$ is also updated by Equation (15):

$$\hat{\lambda}^{n+1} \leftarrow \hat{\lambda}^n + \tau \left(\hat{f} - \sum_k \hat{u}_k^{n+1} \right) \quad (15)$$

The updating stops until following equation is set up,

$$\sum_k \|\hat{u}_k^{n+1} - \hat{u}_k^n\|_2^2 / \|\hat{u}_k^n\|_2^2 < \varepsilon \quad (16)$$

From the above, the first IMF imf_1 is the baseline drift noise.

3.2. Proposed Adaptive Decomposition

In order to get more reasonable decomposition results, we define $f_{new} = f - imf_1$. Document 19 proposed that the penalty factor introduced by the traditional VMD method is inversely proportional to the noise level in the signal. In a limited high frequency noise environment, we can assume that the weight of penalty is directly proportional to the power (which may be obtained by calculating variance) of each IMF. In order to further improve the adaptability of the penalty factor, we propose $\alpha_k = c_2 \cdot D[u_k(t)]$. From the foregoing description, we can see that the low-frequency signal component has a large penalty factor, which can achieve low-frequency refinement and degrade the noise component in each signal mode. The original augmented Lagrange equation $\mathcal{L}$ becomes:

$$\begin{aligned} \mathcal{L}(\{u_k\},\{\omega_k\},\lambda(t)) &= \alpha_k \sum_{k=1}^K \left\| \partial_t \left[\left(\delta(t) + \frac{j}{\pi t} \right) * u_k(t) \right] e^{-j\omega_k t} \right\|_2^2 + \\ &\left\| f_{new}(t) - \sum_{k=1}^K u_k(t) \right\|_2^2 + \langle \lambda(t), f_{new}(t) - \sum_{k=1}^K u_k(t) \rangle \end{aligned} \quad (17)$$

Since the iterative solution process for each IMF component is performed in the frequency domain, the derivation of the penalty factor requires the use of a time-domain representation of each component. In order to obtain accurate decomposition results, we need to solve the penalty factors for each iteration, and then substitute the new penalty factors into the next iteration, which leads to the loop, which is very time-consuming. As such, we need to unify the derivation process into the frequency domain. Expanding the formula to solve the variance creates the following:

$$\begin{aligned} \alpha_k &= \tfrac{c_2}{N}[(u_{k1} - \overline{u_k})^2 + (u_{k2} - \overline{u_k})^2 + \cdots + (u_{kN} - \overline{u_k})^2] \\ &= \tfrac{c_2}{N}\left(u_{k1}^2 + u_{k2}^2 + \cdots + u_{kN}^2\right) - \tfrac{1}{N^2}(u_{k1} + u_{k2} + \cdots + u_{kN})^2 \end{aligned} \quad (18)$$

According to Parseval's Theorem, we know that $\int |u_k(t)|^2 dt = \int |\hat{u}_k(\omega)|^2 d\omega$ and the first item in the above formula can thus be converted into a frequency representation. By applying the Fourier transform of the second term to the above formula, we can get the frequency–domain description of penalty factor:

$$\alpha_k = c_2 \frac{N \int_0^\infty |\hat{u}_k(\omega)|^2 d\omega - 2\pi \hat{u}_k^2(0)}{2\pi N^2} \qquad (19)$$

where c_2 is a constant and N is the length of the data. Combine the penalty factor into the previous update formulas, we can get the new update formulas for $\hat{u}_k(\omega)$, ω_k and α_k:

$$\begin{cases} \hat{u}_k^{n+1}(\omega) = \frac{\hat{f}_{new}(\omega) - \sum_{i<k} \hat{u}_i^{n+1}(\omega) - \sum_{i>k} \hat{u}_i^n(\omega) + \frac{\lambda^n(\omega)}{2}}{1 + 2\alpha_k^n (\omega - \omega_k^n)^2} \\ \omega_k^{n+1} = \frac{\int_0^\infty \omega |\hat{u}_k(\omega)|^2 d\omega}{\int_0^\infty |\hat{u}_k(\omega)|^2 d\omega} \\ \alpha_k^{n+1} = c_2 \frac{N \int_0^\infty |\hat{u}_k(\omega)|^2 d\omega - 2\pi \hat{u}_k^2(0)}{2\pi N^2} \end{cases} \qquad (20)$$

The Lagrange multiplier update formula and iteration stop criterion remain unchanged.

3.3. Iterative Thresholding and Improved VMD Method

In this section, we obtain a new denoising process by combining the interval threshold and the improved variational mode decomposition. In this paper, the hard and soft thresholding methods, with multiples of Donoho-Johnstone threshold known as universal threshold parameters, are proposed, in order to cut off each IMF after performing the improved VMD. The threshold parameters are defined as:

$$T_k = C \frac{median(|u_k(t) - median(u_k(t))|)}{0.6745} \sqrt{2 \ln N} \qquad (21)$$

where C is a constant and N is the length of the data. It is necessary to adopt a scale dependent threshold for each IMF instead of an identical universal threshold for all IMFs. The values of the scaling factor C are the range [0.6, 1.2]. We use the interval thresholding (IT) method to alleviate the catastrophic consequences caused by the direct application of thresholding. In this method, the first step is to find the zero points of each IMF. The second step is to compare the threshold and the extremum between two zero crossing intervals; if the extremum exceeds the threshold, it will allow all of the samples within the interval to be retained. The interval thresholding can be represented as:

$$\hat{c}_k\left(Z_i^k\right) = \begin{cases} c_k\left(Z_i^k\right), & \left|c_k\left(r_i^k\right)\right| > T_k \\ 0 & \left|c_k\left(r_i^k\right)\right| \leq T_k \end{cases} \qquad (22)$$

where i varies from 1 to N_k, N_k indicates the number of zero crossings of the kth IMF, $c_k\left(r_i^k\right)$ indicates the sample at the time instance r_i^k between the two successive zero points at Z_i^k and Z_{i+1}^k, and $c_k\left(Z_i^k\right)$ refers to the samples from the instant Z_i^k to Z_{i+1}^k of the kth IMF.

Summarizing this denoising process as follows: First, the MCG signal with noise is decomposed into corresponding IMFs by improved VMD. The first order IMF, including the baseline drift noise, is eliminated. Second, the interval thresholding is performed on the rest IMFs. Clearly, the higher the order, the greater the frequency of the IMF. The detailed procedures are as follows:

1. Acquire the MCG signal with noise and initialize the number of modes k, the default of the penalty factor α is 2000, the default of the bandwidth τ is 0.
2. Determine the value of penalty factor by performing the improved VMD method based on the correlation coefficient and obtain the first IMF.
3. Eliminate the first IMF that contains baseline drift noise. Decompose the rest signal with the final update formulas and obtain several IMFs.
4. Perform the interval thresholding operation on the IMFs obtained from step 3.
5. Add all the processed IMFs together and refactor the MCG signal.

In Figure 1 we show the flow chart of the improved VMD algorithm:

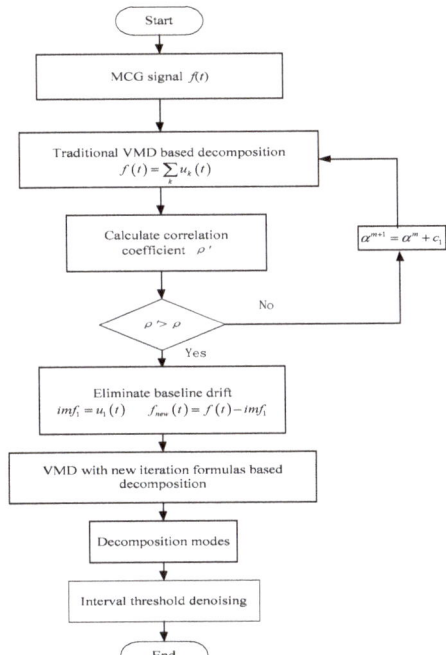

Figure 1. The flow chart of the improved Variational Mode Decomposition (VMD) algorithm.

4. Results and Discussion

Three different types of noise have been added to the MCG signal, in order to investigate the effectiveness of denoising by an improved VMD and the interval thresholding method. The types of noise include a low frequency (0.3 Hz) sinusoidal signal for simulating the baseline drift, 50 Hz sinusoidal signal for simulating the interference at power line frequency, and high frequency random noise. In Figure 2, we compare the waveforms of an original simulation signal and a mixed signal. It is clearly seen that background noise affects signal analysis. It is very significant in the process of measuring MCG signals to detect heart disease.

The signal is denoised by different algorithms. In Figure 3a,b, we show the signal decomposition results obtained by the EEMD method. In Figure 3, according to the length of the data, the MCG signal is decomposed into ten IMFs, and the frequency of IMFs reduces as the order increases. In general, the low frequency smooth variation of the baseline is expected to be contained in the residue of the higher order IMFs. However, we cannot determine the low-frequency component that contains baseline drift noise. It is evident that we can extract signal characteristics from IMF2 to IMF6. Unfortunately, other IMFs may also contain useful signal components that are invisible to the human eye. Given this, we apply thresholds only to a few low order IMFs that contain contributions from the high frequency noise components, then exclude a few high order IMFs that contain the low frequency contents with a view, to ensure that the low frequency contents of the MCG signal are not affected or distorted by thresholding.

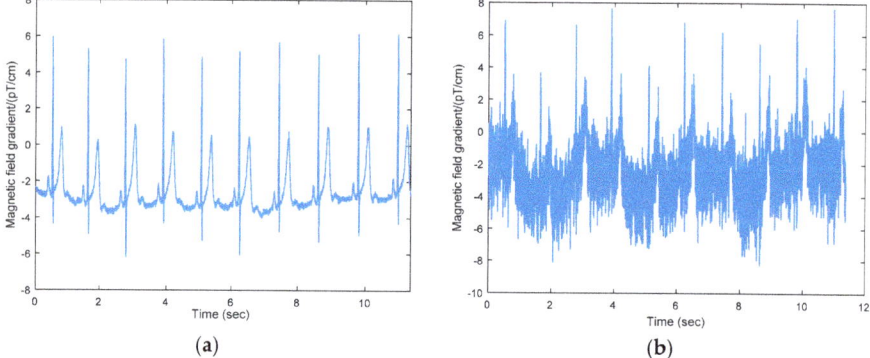

Figure 2. A comparison of the approximately pure magnetocardiography (MCG) signal with the noisy signal; waveform characteristics can be clearly seen in (**a**); (**b**) shows the signal waveform with three kinds of noise. The signal is drowned in the noise under high noise circumstances.

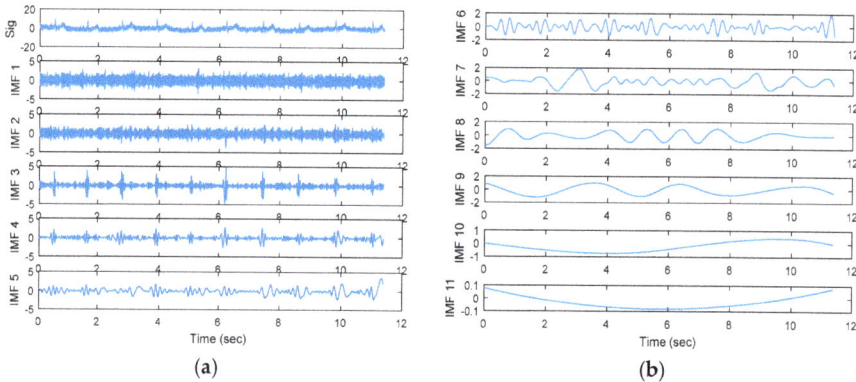

Figure 3. The noisy signal and the intrinsic mode functions (IMFs) obtained by Ensemble Empirical Mode Decompositioning (EEMD) are shown in (**a**,**b**).

In Figure 4, we compare the original signal with the reconstructed signals obtained from EEMD based denoising methods using soft and hard thresholding. The results show that the hard threshold processing can reconstruct the QRS peak waves, but there are obvious errors in the reconstruction of other signal parts. The signal obtained by soft thresholding processing has errors compared with the original signal, especially the QRS peaks. It is seen from Figure 4 that EEMD algorithm is difficult to distinguish noise components from signal components. And the number of low frequency IMFs is too many to result in waveform distortion after performing thresholding operation. It may also be noted that, hard-IT (hard thresholding subsequently interval thresholding) is adopted for EEMD based denoising of the experimental data. At the same time, we can see that the signal waveform is not smooth and slightly distorted.

Based on the bandwidth of the measurement signal and multiple tests, the number of modes decomposed by VMD is assigned to 6. The original algorithm proposer studied some of the convergence characteristics of the VMD algorithm and its sensitivity to the initial conditions, then got relatively suitable initialization parameters. The initial value of quadratic penalty α is assigned to 2000, and the default of the bandwidth τ is 0. With this method, the MCG signal is divided into several frequency bands centering on respective center frequency, which benefits subsequent operation. As shown in Figure 5a, the MCG signal is divided into six IMFs. The baseline drift noise is found in IMF1. In Figure 5b, we compare the original signal with the reconstructed signals obtained from VMD methods with soft and hard thresholding. The results show that the baseline of the reconstructed signal is not uniform with the original signal. To make matters worse, there is serious distortion in the reconstructed signal from the soft thresholding processing. It is obvious that the algorithm cannot effectively remove baseline drift noise. The decomposition process of the algorithm lacks of adaptability. Improper decomposition results can easily lead to waveform distortion after threshold processing.

To get a better result, we use the proposed method to decompose the signal. According to many experiments, we choose the parameter value with better effect as the next simulation initial values. The threshold value ρ of correlation coefficient is assigned as 0.95. The value of the scaling factor c_1 is the range 1500 and 2000. The value of c_2 can be set according to the detailed data features and is assigned as 3.5. As Figure 6 shown, the baseline noise is extracted. After removing the noise, the signal is divided into six modes adaptively. According to iterative formulas for multiple iterations, the penalty factors for six components are [1790, 987, 956, 218, 223, 208]. The decomposition results are shown in Figure 7a and the denoising results are shown in Figure 7b.

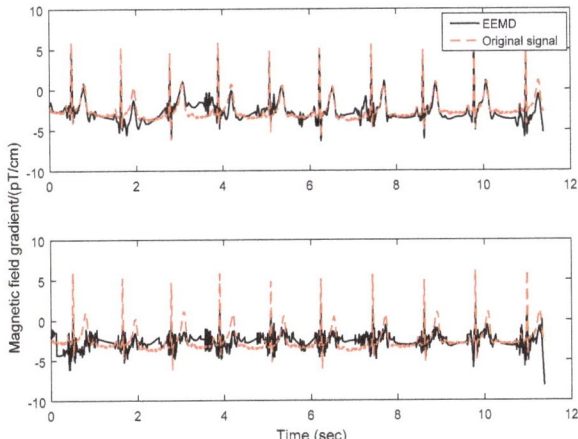

Figure 4. A comparison of the original signal (dotted line) with the reconstructed signals (solid line) obtained from EEMD based denoising methods with soft and hard thresholding. The panel above is the result of hard thresholding processing and the following panel is the result of soft thresholding processing.

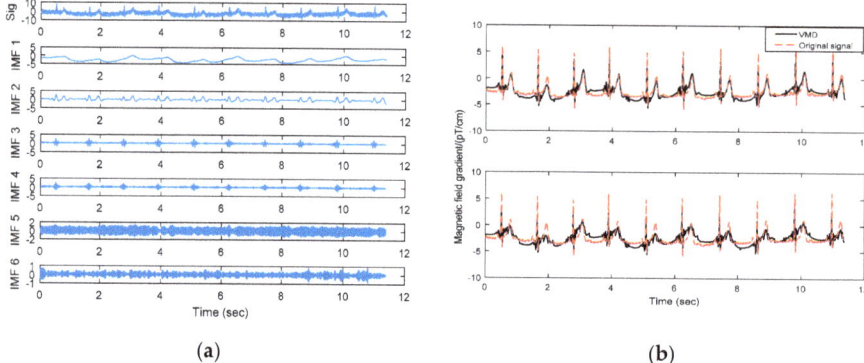

(a)   (b)

Figure 5. The noisy signal and the IMFs obtained from VMD are shown in (**a**). A comparison of the original signal (dotted line) with the reconstructed signals (solid line) obtained from VMD based denoising methods with soft and hard thresholding is shown in (**b**). The panel above is the result of hard thresholding processing and the following panel is the result of soft thresholding processing.

The range of the value of the penalty factors can support the adaptive decomposition result. From this result, it can be seen that the low-frequency IMF's penalty factors are large, and the penalty factor increases approximately with the increasing of the center frequency, achieving the purpose of the meticulous decomposition of low-frequency signal component. In addition, we can obtain a set of different penalty factors by adjusting the size of c_2 to adjust the decomposition result. By comparing the reconstructed signal with the original signal, we can find that the fitting degree between the reconstructed signal and the original signal is good, and all three kinds of noise in the signal are effectively removed. It may also be noted that, signal to noise ratio improvement is much better of the hard interval thresholding (IT) method compared to the soft interval thresholding.

In order to better compare the performance of the algorithms, we use the root-mean-square error (the square root of the mean of the sum of squared residuals, RMSE) to characterize the fitting degree of the reconstructed signal and the original signal. In Figure 8, we compare the root-mean-square deviation (RMSE) of the three methods with the input signal-to-noise ratio (SNR). In the case of low input SNR, the RMSE of the improved VMD method is significantly less than the other two methods, and the method has better denoising performance even with the low input SNR.

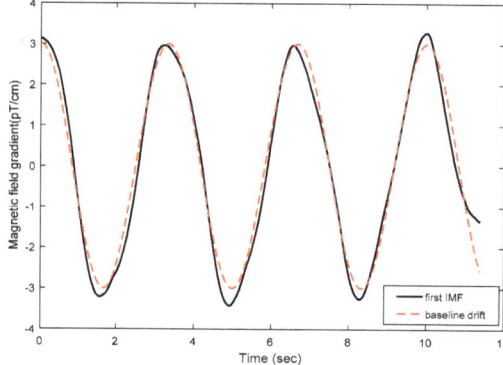

Figure 6. The first IMF obtained from the improved VMD method (solid line) and baseline drift noise (dotted line). It is seen that the solid line and the dotted line basically coincide. This method can effectively remove baseline drift noise.

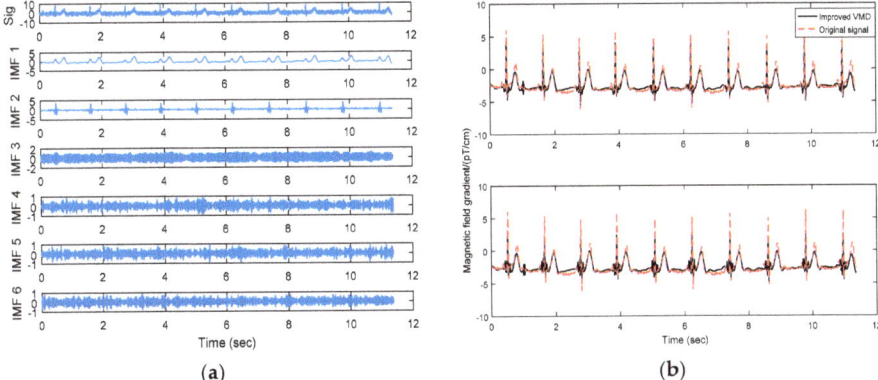

Figure 7. (a) The noisy signal and the IMFs obtained from the improved VMD method; (b) A comparison of the original signal (dotted line) with the reconstructed signals (solid line) obtained from the improved VMD method with soft and hard thresholding for interval thresholding. The panel above is the result of hard thresholding processing and the following panel is the result of soft thresholding processing.

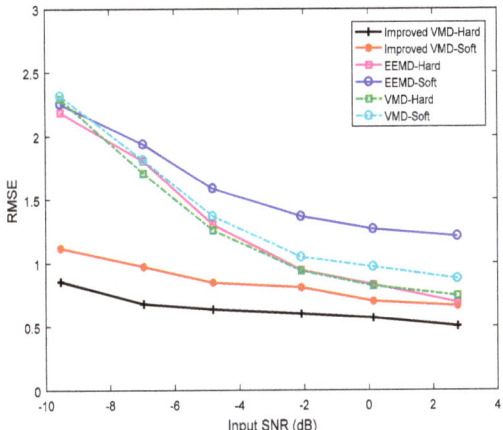

Figure 8. The root-mean-square deviation (RMSE) of the EEMD, the VMD, and the improved VMD methods are revealed, for both soft and hard thresholding of interval thresholding. The improved VMD method outperforms other methods.

In Figure 9, we compare the performance of the original VMD method, the improved VMD method, and the EEMD methods, using soft and hard thresholding in each case. For computing SNR, the logarithmic ratio of variance of a signal (from the beginning of P-wave to the end of T-wave for one cardiac cycle) to the variance of noise (from the end of T-wave to the beginning of P-wave, i.e., in the TP interval) has been taken. The reduction in the SNR for the soft thresholding method is due to the reduction of signal components of the lower frequency IMFs. Hereafter, hard-IT (hard thresholding is subsequently applied to the interval thresholds) is appropriate for both EEMD and VMD based denoising of the experimental data. It is seen from Figure 9 that the improved VMD method is capable of achieving better SNR when compared with EEMD and VMD methods.

From the above simulation results, we can see that the proposed algorithms have better denoising performance compared with EEMD and VMD methods. It should be noted, however, that the waveform is still distorted, even by the proposed method; this is a crucial detail in electrocardiographic-like signals. One encouraging factor, is that results of medical research have shown that the QRS spikes and S-T waves contain information on the main electrical function parameters of the heart. The MCG signal is filtered by the algorithm proposed in this paper, in order to obtain QRS spikes and S-T waves that approximate the original signal. Although there are slight disturbances in the QRS spikes and S-T waves obtained by filtering, it does not affect the calculation of heart related parameters! Measurements of magnetic field energy and current density remain accurate). In the case of the low input signal-to-noise ratio used in this paper, the SNR improvement of the proposed algorithm can be up to 20 dB. The algorithm filtering results can support feature extraction of MCG and detection of heart disease.

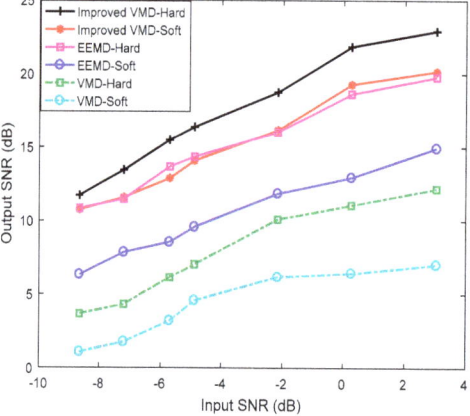

Figure 9. The variation of the output signal-to-noise ratio (SNR) by EEMD, VMD and the improved VMD methods with soft and hard thresholds for interval thresholding. The improved VMD method outperforms other methods.

Despite these positive results, the MCG signals measured in a clinical non-magnetic shielding environment would still contain a large amount of background noise, which would still cause the waveform, denoised by the improved VMD method proposed in this paper, to produce severe distortion. In order to solve this problem, we need to study the spatial filtering technology in order to further denoise the MCG signal, in efforts to achieve the purpose of joint filtering in the time domain, frequency domain, and space domain.

5. Conclusions

The proposed method in this paper overcomes the unadaptable quadratic penalty problem of VMD, which improves the availability and precision of denoising of the MCG signal. This method adaptively adjusts the bandwidth of modes by repeatedly executing the VMD method with different quadratic penalties. The low-frequency noise is eliminated, according to the correlation coefficient for baseline drift noise, and the first mode. The new update formulas are used to decompose residual signals adaptively. Then, threshold processing is performed on each IMF to eliminate other noise. The simulation experiments show the superiority of the improved VMD in denoising performance of the MCG signal. The acceleration of the proposed method, and the suitable signal preprocessing methods, should be considered in future research and applications.

Author Contributions: Conceptualization, Y.L. and Q.G.; Methodology, C.H.; Software, C.H.; Validation, Y.L. and Q.G.; Writing—original draft, C.H.; Writing—review & editing, Y.L. and C.H.

Funding: This work is supported partially by the National Key Research and Development Program of China (2016YFC0101700), by National Natural Science Foundation of China 61301201 and 61371175, Heilongjiang Postdoctoral Research Foundation LBH-Q14039.

Acknowledgments: The authors are grateful to the anonymous referees for their valuable comments and suggestions that improved this paper.

Conflicts of Interest: The authors declare no conflicts of interest.

References

1. Mäkijärvi, M.; Montonen, J.; Toivonen, L. Identification of patients with ventricular tachycardia after myocardial infarction by high-resolution magnetocardiography and electrocardiography. *J. Electrocardiol.* **1993**, *26*, 117–124. [CrossRef]
2. Shanehsazzadeh, F.; Fardmanesh, M. Low Noise Active Shield for SQUID-Based Magnetocardiography Systems. *IEEE Trans. Appl. Supercond.* **2017**, *99*, 1–5. [CrossRef]
3. Farré, J.; Shenasa, M. Medical Education in Electrocardiography. *J. Electrocardiol.* **2017**, *50*, 400–401. [CrossRef] [PubMed]
4. Fenici, R.; Brisinda, D. Magnetocardiography provides non-invasive three-dimensional electroanatomical imaging of cardiac electrophysiology. *Anatol. J. Cardiol.* **2006**, *22*, 595. [CrossRef] [PubMed]
5. Ha, T.; Kim, K.; Lim, S. Three-Dimensional Reconstruction of a Cardiac Outline by Magnetocardiography. *IEEE Trans. Biomed. Eng.* **2015**, *62*, 60–69. [CrossRef] [PubMed]
6. Mariyappa, N.; Parasakthi, C.; Sengottuvel, S. Dipole location using SQUID based measurements: Application to magnetocardiography. *Phys. C Supercond.* **2012**, *477*, 15–19. [CrossRef]
7. Tiporlini, V.; Alameh, K. Optical Magnetometer Employing Adaptive Noise Cancellation for Unshielded Magnetocardiography. *Univ. J. Biomed. Eng.* **2013**, *1*, 16–21. [CrossRef]
8. Kim, K.; Lee, Y.H.; Kwon, H. Averaging algorithm based on data statistics in magnetocardiography. *Neurol. Clin. Neurophysiol.* **2004**, *2004*, 42. [PubMed]
9. Dang-Ting, L.; Ye, T.; Yu-Feng, R.; Hong-Wei, Y.; Li-Hua, Z.; Qian-Sheng, Y.; Geng-Hua, C. A novel filter scheme of data processing for SQUID-based Magnetocardiogram. *Chin. Phys. Lett.* **2008**, *25*, 2714–2717. [CrossRef]
10. Tiporlini, V.; Nguyen, N.; Alameh, K. Adaptive noise canceller for magnetocardiography. In Proceedings of the High Capacity Optical Networks and Enabling Technologies, Riyadh, Saudi Arabia, 19–21 December 2011; pp. 359–363.
11. Dong, Y.; Shi, H.; Luo, J. Application of Wavelet Transform in MCG-signal Denoising. *Mod. Appl. Sci.* **2010**, *4*, 20. [CrossRef]
12. Huang, N.E.; Shen, Z.; Long, S.R.; Wu, M.C.; Shih, H.H.; Zheng, Q.; Yen, N.C.; Tung, C.C.; Liu, H.H. The empirical mode decomposition and the Hilbert spectrum for nonlinear and non-stationary time series analysis. *Proc. R. Soc. Lond. A* **1998**, *454*, 903–995. [CrossRef]
13. Attoh-Okine, N.; Barner, K.; Bentil, D.; Zhang, R. Editorial: The empirical mode decomposition and the hilbert-huang transform. *EURASIP J. Adv. Signal Process.* **2008**. [CrossRef]
14. Wu, Z.; Huang, N.E. Ensemble empirical mode decomposition method: A noise-assisted data analysis method. *Adv. Adapt. Data Anal.* **2009**, *1*, 1–41. [CrossRef]
15. Tong, W.; Zhang, M.; Yu, Q. Comparing the applications of EMD and EEMD on time–frequency analysis of seismic signal. *J. Appl. Geoph.* **2012**, *83*, 29–34.
16. Mariyappa, N.; Sengottuvel, S.; Parasakthi, C. Baseline drift removal and denoising of MCG data using EEMD: Role of noise amplitude and the thresholding effect. *Med. Eng. Phys.* **2014**, *36*, 1266–1276. [CrossRef] [PubMed]
17. Mariyappa, N.; Sengottuvel, S.; Patel, R. Denoising of multichannel MCG data by the combination of EEMD and ICA and its effect on the pseudo current density maps. *Biomed. Signal Process. Control.* **2015**, *18*, 204–213. [CrossRef]
18. Smruthy, A.; Suchetha, M. Real-Time Classification of Healthy and Apnea Subjects Using ECG Signals With Variational Mode Decomposition. *IEEE Sens. J.* **2017**, *17*, 3092–3099. [CrossRef]

19. Dragomiretskiy, K.; Zosso, D. Variational Mode Decomposition. *IEEE Trans. Signal Process.* **2014**, *62*, 531–544. [CrossRef]
20. Sun, Z.G.; Lei, Y.; Wang, J. An ECG signal analysis and prediction method combined with VMD and neural network. In Proceedings of the IEEE International Conference on Electronics Information and Emergency Communication, Macau, China, 21–23 July 2017; pp. 199–202.
21. Maji, U.; Pal, S. Empirical mode decomposition vs. variational mode decomposition on ECG signal processing: A comparative study. In Proceedings of the International Conference on Advances in Computing, Communications and Informatics, Jaipur, India, 21–24 September 2016; pp. 1129–1134.
22. Mert, A. ECG signal analysis based on variational mode decomposition and bandwidth property. In Proceedings of the IEEE Signal Processing and Communication Application Conference, Zonguldak, Turkey, 16–19 May 2016; pp. 1205–1208.
23. Viswanath, A.; Jose, K.J.; Krishnan, N. Spike Detection of Disturbed Power Signal Using VMD ☆. *Procedia Comput. Sci.* **2015**, *46*, 1087–1094. [CrossRef]
24. Kopsinis, Y.; Mclaughlin, S. Development of EMD-Based Denoising Methods Inspired by Wavelet Thresholding. *IEEE Trans. Signal Process.* **2009**, *57*, 1351–1362. [CrossRef]
25. Daubechies, I.; Lu, J.; Wu, H.T. Synchrosqueezed wavelet transforms: An empirical mode decomposition-like tool. *Appl. Comput. Harmonic Anal.* **2011**, *30*, 243–261. [CrossRef]
26. Gilles, J. Empirical Wavelet Transform. *IEEE Trans. Signal Process.* **2013**, *61*, 3999–4010. [CrossRef]
27. Unser, M.; Sage, D.; Van, D.V.D. Multiresolution monogenic signal analysis using the Riesz-Laplace wavelet transform. *IEEE Trans. Image Process.* **2009**, *18*, 2402–2418. [CrossRef] [PubMed]
28. Bertsekas, D.P. Multiplier methods: A survey. *Automatica* **1976**, *12*, 133–145. [CrossRef]
29. Hestenes, M.R. Multiplier and gradient methods. *J. Optim. Theory Appl.* **1969**, *4*, 303–320. [CrossRef]
30. Rockafellar, R.T. A dual approach to solving nonlinear programming problems by unconstrained optimization. *Math. Program.* **1973**, *5*, 354–373. [CrossRef]
31. Boley, D. Local Linear Convergence of the Alternating Direction Method of Multipliers on Quadratic or Linear Programs. *Siam J. Optim.* **2013**, *23*, 2183–2207. [CrossRef]

© 2018 by the authors. Licensee MDPI, Basel, Switzerland. This article is an open access article distributed under the terms and conditions of the Creative Commons Attribution (CC BY) license (http://creativecommons.org/licenses/by/4.0/).

Article

Magnetic and Structural Properties of Barium Hexaferrite Nanoparticles Doped with Titanium

Abdul Raouf Al Dairy [1], Lina A. Al-Hmoud [2,*] and Heba A. Khatatbeh [1]

[1] Department of Physics, Yarmouk University, Irbid 21163, Jordan; abedali@yu.edu.jo (A.R.A.D.) heba.haboush92@yahoo.com (H.A.K.)
[2] Department of Electrical Power Engineering, Yarmouk University, Irbid 21163, Jordan
* Correspondence: lina.hmoud@yu.edu.jo; Tel.: +962-7-9018-4875

Received: 6 May 2019; Accepted: 24 May 2019; Published: 28 May 2019

Abstract: Samples of Barium Hexaferrite doped with Titanium $BaFe_{12-x}Ti_xO_{19}$ with (x = 0.0, 0.2, 0.4, 0.6, 0.8, 1.0) were synthesized by the sol–gel auto-combustion technique. The powdered samples were divided into two parts, one sintered at 850 °C and another sintered 1000 °C for 1 h and samples were characterized by different experimental techniques. The XRD patterns confirmed the presence of M-type hexaferrite phase. The sizes of the crystallites were calculated by the Scherer equation, and the sizes were in the range of 27–42 nm. Using the hysteresis loops, the saturation magnetization M_s, remanence (M_r), the relative ratio (M_r/M_s), and the coercivity (H_c) were calculated. The study showed that the saturation magnetization (M_s) and remanence (M_r) decreased with increasing titanium concentration and were in the range from 44.65–17.17 emu/g and 23.1–7.7 emu/g, respectively. The coercivity (H_c) ranged between 0.583 and 4.51 (kOe). The magnetic properties of these Barium Hexaferrite doped with Titanium indicated that they could be used in the recording equipment and permanent magnets.

Keywords: Barium hexaferrite; titanium; hysteresis; X-ray diffraction; permanent magnet applications

---

1. Introduction

There are vast numbers of applications based on magnetic materials and as a result of that; our lives have improved. They are used in the fabrication of many types of equipment and have a major role in the advancement of technological and industrial products. The types of magnetic materials, according to their magnetic behavior, are one of five types; Diamagnetic, Paramagnetic, Ferromagnetic, Antiferromagnetic, and Ferrimagnetic materials [1–3]. Magnetic ferrites, first discovered in the 1950s, are ceramics made from iron oxides with one or more additional metals chemically added [4,5]. These ferrites had been considered very highly valuable electronic materials for many decades. The ferrite compounds have a cubic structure, but there is also a group of ferrites with a hexagonal crystal structure, known as hexaferrites [6]. There was an increasing degree of interest in hexaferrites, and it is still growing today. They have been massively produced and became important materials; commercially and technologically used in many electrical systems, such as permanent magnets, magnetic recording and data storage devices [6,7].

The hexaferrites are complex oxide systems with a general chemical formula $AO–Fe_2O_3–MO$, where A is a large divalent cation, i.e., Ba, Sr, Ca, and M are a small divalent cation, i.e., Mn, Fe, Co, Ni, Cu, and Zn. They can be classified on the basis of their chemical composition and the A–M combination and the crystal structure into six fundamental, structural types: M, W, Y, X, U, and Z [8–13].

The ferromagnetic and ferromagnetic types show nonlinear relation between the magnetization M of the compound and the applied H. The hysteresis loop depicts the behavior of the magnetization M of the sample with the variation of the applied field H. As H increases, the magnetization increases up to its highest value; this defines an important characteristic parameter of the material called the saturation magnetization M_s at high enough applied fields. The value of the field needed to reach the saturation magnetization depends on the nature of the material, method of preparation, and other intrinsic and extrinsic parameters. The magnetization retains a value even at zero applied fields called the remnant magnetization or the remanence M_r. This parameter has particular importance in the case of permanent magnets production because it defines the magnetization of a magnet in the absence of an applied external field. The size of the opposite field needed to reduce M to zero is called the coercivity H_c. The value of the coercive field, the coercivity, defines the so-called magnetic hardness of the magnetic material [14].

Many research groups prepared these ferrites and investigated the effect of the dopants on the magnetic properties of such M-type magnetic hexaferrites [15–33]. The hexaferrites are of great interest for applications in the microwave technology and others. S.V. Trukhanov et al. studied the effect of gallium doping on the properties of barium hexaferrite, $BaFe_{12-x}Ga_xO_{19}$ ($x \leq 1.2$), prepared by the ceramic technology method [34,35]. They showed that the unit cell monotonically decreases with increasing x and these Ga doped hexaferrites can effectively absorb high-frequency electromagnetic radiation. The maximum of the real part of permeability depends on the level of substitution by titanium cations and it is located in the region of 5–6 GHz. For gallium-substituted hexaferrites, the real part of permittivity decreases more slowly at low frequencies and almost monotonically with concentration. In last case, the real and imaginary parts of the permeability have a peak in the region of 49–51 GHz, which is determined by the level of diamagnetic substitution. Researchers have prepared $(BaFe_{11.9}Al_{0.1}O_{19})_{1-x}(BaTiO_3)_x$ with $x = 0. \ 0.25, 0.5, 0.75$, and 1 bicomponent compounds using the ceramic technique [36]. The researchers reported that these hexaferrites exhibited ferroelectricity at room temperature and the coercive field was lower due to the contribution of the microstructure-dependent shape anisotropy to the magnetic anisotropy energy and the behavior of these samples was discussed based on the grain size, density, and porosity. The magnetic and dipole moments in $BaFe_{12-x}In_xO_{19}$ hexaferrites were studied by S. V. Trukhanov group [37]. These indium doped samples were prepared by solid reaction method as published by coworkers of the same group before [38]. The samples were studied by high resolution neutron powder diffraction and vibrating sample magnetometry in the temperature range of 4–730 K. They showed that spontaneous polarization was established due to the displacement of Fe^{3+}(In^{3+}) cations and the appearance of nonzero electric dipole moment, which causes the formation of the z-component of the spontaneous polarization. The crystal structure and magnetic properties of $BaFe_{12-x}Me_xO_{19}$ (Me = In^{3+} or Ga^{3+} and $x = 0.1$–1.2) solid solutions were studied using the time–of–flight neutron diffraction method [39]. The workers reported that the electric field–induced polarization was observed in these barium hexaferrite solid solutions at room temperature. Using the Gorter's model, the researchers found that the magnetic moments of iron ions were oriented along the hexagonal axis which is the easy axis of magnetization. The previously mentioned works of S.V. Trukhanov and co-workers proved that the magnetoelectric characteristics of M-type hexaferrites fabricated by a modified ceramic technique can be more advanced than those for the well-known room temperature $BiFeO_3$ orthoferrite multiferroic.

The main goal of this project was to study the effect of Titanium substitution on the magnetic and structural properties of the barium hexaferrites prepared according to the formula $BaFe_{11.9}Ti_xO_{19}$ with $x = 0.0, 0.2, 0.4, 0.6, 0.8, 1.0$.

2. Experimental Techniques

A number of ferrite samples $BaFe_{12-x}Ti_xO_{19}$ with ($x = 0.0, 0.2, 0.4, 0.6, 0.8, 1.0$) were prepared by using the sol-gel auto combustion technique. The chemicals used for the preparation of the samples were $Fe(NO_3)_3$, $Ba(NO_3)_2$, and $TiCl_3$ and were dissolved in 100 ml of de-ionized water. Citric acid was added to the solution, and the molar ratio for the metallic mixture to acid was kept at 3:1 [33]. Using a hot plate and continuous stirring, the solution was heated up to a temperature of 80 °C. A solution of ammonia was added to the mixture to reach a pH value of 8.0. The solution was then heated slowly to 450 °C for several hours until a viscous brown gel mixture is formed. The gel was heated further to a temperature of 550 °C and the gel ignited and was burned out completely and formed a fluffy brown colored powder. This powder was divided into two samples, using a furnace one sample was sintered at 850 °C and the second sample was sintered at 1000 °C each for a period of one hour.

The structure of the samples was studied using X-ray powder diffraction (with Cu-K_α radiation λ = 1.5405 Å). Infrared spectroscopy was used to ensure the formation of M-type hexaferrite phase. The scanning electron microscope was used to study the microstructure of samples. Finally, the magnetic properties were studied at room temperature using a Vibrating Sample Magnetometer.

3. Results and Discussion

Figure 1 shows the XRD pattern for $BaFe_{12}O_{19}$ and the XRD patterns for doped $BaFe_{12-x}Ti_xO_{19}$ samples sintered at 850 °C, while Figure 2 shows the XRD pattern for $BaFe_{12}O_{19}$ and those patterns for doped $BaFe_{12-x}Ti_xO_{19}$ samples sintered at 1000 °C. It is clear that the XRD pattern for the $BaFe_{12}O_{19}$ (i. e., $x = 0$) as the prepared sample is consistent with the standard pattern (JPCDS #: 00-043-0002) for BaM compound but with a higher percentage of that phase at the higher sintering temperature. This leads to the conclusion that the samples must be sintered at temperatures even higher than 1000 °C. As seen from Figure 2, the patterns indicated the presence of α-Fe_2O_3 in samples sintered at 1000 °C as seen from the peak at 33.1 degree and other lower peaks at 24.2, 49.4, and 54 degree and the content of α-Fe_2O_3 changes with x and was the highest for $x = 0.6$. It is also clear from peaks at 53.2 and 61.8 degree that there are traces of $FeTiO_3$ oxide and the peak at 28.8 degree indicated a presence of a small trace of $BaTiO_3$.

The Scherrer formula was then used to calculate the average crystallite sizes for the samples:

$$D = \frac{k\lambda}{\beta \cos \theta}. \quad (1)$$

Here, D is the average crystallite size, K is called the Scherrer's constant and is taken equal to 0.89 for the hexaferrite, λ is the x-ray wavelength and equals 1.5405 Å, β is the width of the peak at half maximum and is measured in radians, and θ is the position of the peak. Using several peaks, the average crystallite size for all samples was calculated; the results and calculations are summarized in Tables 1 and 2.

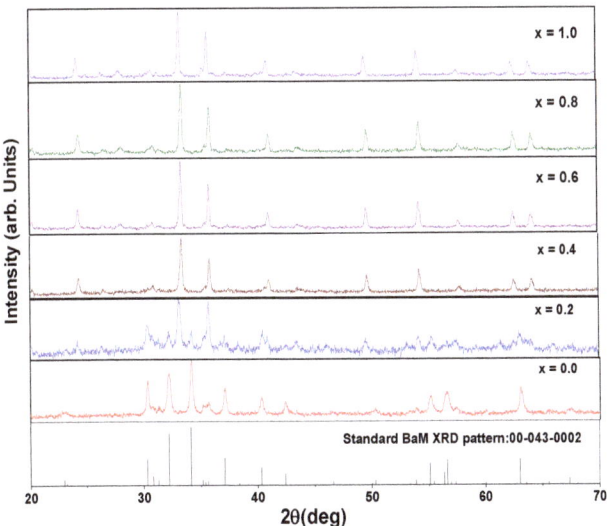

Figure 1. XRD patterns for $BaFe_{12-x}Ti_xO_{19}$ samples with (x = 0.0–1.0) sintered at 850 °C for 1 h and shown the standard pattern of $BaFe_{12}O_{19}$ (file no.: 043-0002).

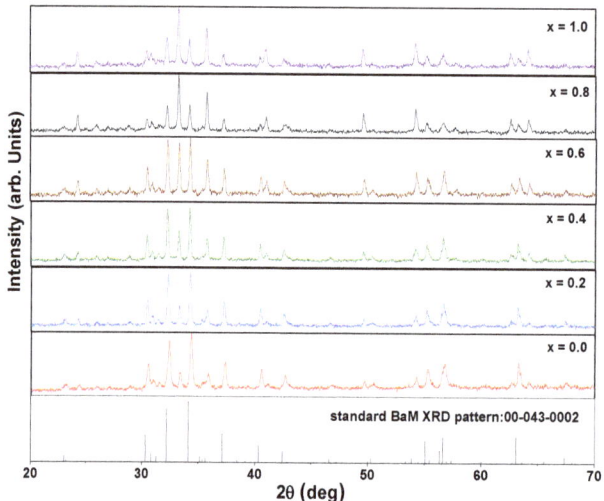

Figure 2. XRD patterns BaFe$_{12-x}$Ti$_x$O$_{19}$ samples with (x = 0.0–1.0) sintered at 1000 °C for 1 h and shown the standard XRD pattern of BaFe$_{12}$O$_{19}$ (file no.: 043-0002).

The average crystallite sizes were found to be in the range of (27–36) nm for samples sintered at 850 °C, and the average crystallite sizes for samples sintered at 1000 °C were found to be in the range of (35–41.5) nm. This may lead us to conclude that Ti substitution improved the BaM phase crystallinity especially at the sintering temperature of 1000 °C.

Table 1. The calculated average crystallite sizes for the system BaFe$_{12-x}$Ti$_x$O$_{19}$ with (x = 0.0–1.0) sintered at T = 850 °C.

Sample	(2θ)-Position (Degrees)	Size (nm)	Average Size (nm)
BaFe$_{12}$O$_{19}$ (x = 0.0)	32.19	24.8	27.2
	34.14	29.3	
	37.1	29.6	
	40.33	27	
	55.1	25.3	
BaFe$_{11.8}$Ti$_{0.2}$O$_{19}$ (x = 0.2)	33.1	27.3	30.3
	35.67	33	
	49.51	25.4	
	54.1	35.3	
BaFe$_{11.6}$Ti$_{0.4}$O$_{19}$ (x = 0.4)	33.3	34.1	34
	35.8	35.9	
	49.65	33.3	
	54.26	32.7	
BaFe$_{11.4}$Ti$_{0.6}$O$_{19}$ (x = 0.6)	33.31	39	36.5
	35.79	41	
	49.69	34.6	
	54.24	31.5	
BaFe$_{11.2}$Ti$_{0.8}$O$_{19}$ (x = 0.8)	33.39	34.2	34
	35.86	35.9	
	49.7	33.3	
	54.3	32.7	
BaFe$_{11.0}$Ti$_{1.0}$O$_{19}$ (x = 1.0)	33.31	39	36.3
	35.69	41.3	
	49.54	34.6	
	54.15	30.4	

Table 2. The calculated average crystallite sizes for the system BaFe$_{12-x}$Ti$_x$O$_{19}$ with (x = 0.0–1.0) sintered at T = 1000 °C.

Sample	(2θ)-Position (Degrees)	Size (nm)	Average Size (nm)
BaFe$_{12}$O$_{19}$ (x = 0.0)	32.4	30.3	34.9
	34.34	41	
	40.55	36.4	
	55.31	31.7	
BaFe$_{11.8}$Ti$_{0.2}$O$_{19}$ (x = 0.2)	32.29	38.9	38.7
	34.23	41.1	
	40.44	41.9	
	55.1	32.8	
BaFe$_{11.6}$Ti$_{0.4}$O$_{19}$ (x = 0.4)	32.26	43.1	41.5
	34.22	45.7	
	40.42	46.5	
	55.19	30.6	
BaFe$_{11.4}$Ti$_{0.6}$O$_{19}$ (x = 0.6)	32.29	43	41
	34.27	45.7	
	40.48	39.9	
	54.23	35.3	
BaFe$_{11.2}$Ti$_{0.8}$O$_{19}$ (x = 0.8)	33.27	45.6	40.5
	34.24	43.3	
	40.96	36.5	
	54.19	36.7	
BaFe$_{11.0}$Ti$_{1.0}$O$_{19}$ (x = 1.0)	33.23	43.1	38
	35.7	41.3	
	40.93	32.2	
	54.19	35.3	

To prove the presence of M-type hexaferrite phase, Infra-Red (IR) analysis was used because the IR- spectra can point to its presence. The samples were mixed with 0.05% KBr in order to get an acceptable resolution of the compound bands. IR spectra of the as-prepared samples were obtained with a wave number varying between 300 cm^{-1} to 1000 cm^{-1} and the results are summarized in Figures 3 and 4 for the two sintering temperatures, respectively.

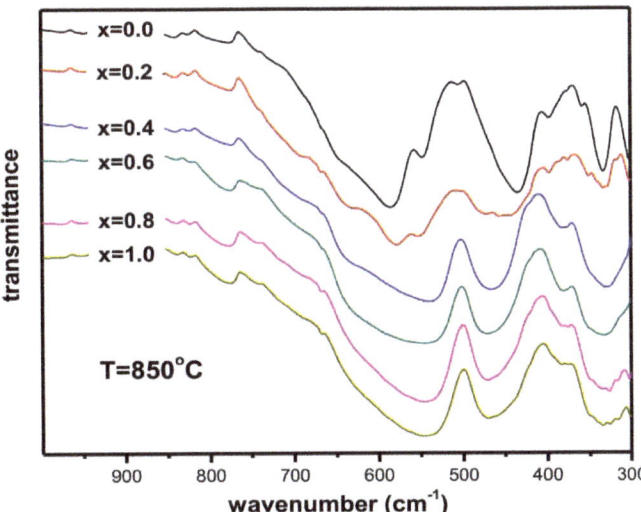

Figure 3. Infra-Red (IR) spectra of BaFe$_{12-x}$Ti$_x$O$_{19}$ sintered at T = 850 °C.

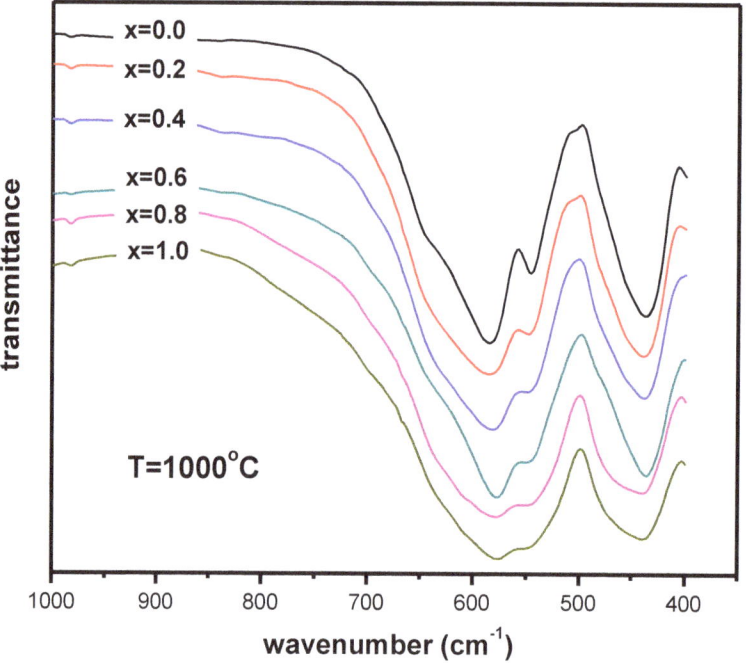

Figure 4. IR spectra of BaFe$_{12-x}$Ti$_x$O$_{19}$ sintered at T = 1000 °C.

As seen from Figures 3 and 4, the IR spectra of the sintered samples have absorption bands at wave numbers in the range 430–590 cm^{-1}, which are characteristic to the formation of ferrites. The peaks were clearer for those samples sintered at 1000 °C. The peak at about 430 cm^{-1} is due to bending of the absorption bands of the metal-oxygen and that at about 590 cm^{-1} is due stretching of these bands.

Figure 5 shows an SEM image for one of the samples, namely BaFe$_{12-x}$Ti$_x$O$_{19}$ (with x = 0.2), which was sintered at T = 1000 °C. It shows that grains are ultrafine and almost have homogeneous distribution. The image also shows that the platelets have sizes in the range of 200 nm to 500 nm.

Figure 5. SEM image for BaFe$_{12-x}$Ti$_x$O$_{19}$ (x = 0.2) sample sintered at T = 1000 °C.

Based on X-ray results and IR data, we believe that samples are improved if heat-treated at 1000 °C.

We studied only magnetic properties for samples sintered at 1000 °C. Magnetic measurements were carried out at room temperature by a VSM with applied field between 0 and 10 kOe. The group of hysteresis loops (HL) for all samples is shown in Figure 6. The loops do indicate that all the samples are indeed hard magnetic materials. The values of remanence magnetization and coercive field for each sample were read directly from graphs of loops. Since it is expected that samples have high magnetic anisotropy, we used the law of approach to saturation (LAS) to calculate the saturation magnetization. This is simply found by plotting M versus $1/H^2$ for values taken from high field region and all samples gave perfect straight linear graphs. The intercept of that line is equal to the saturation magnetization.

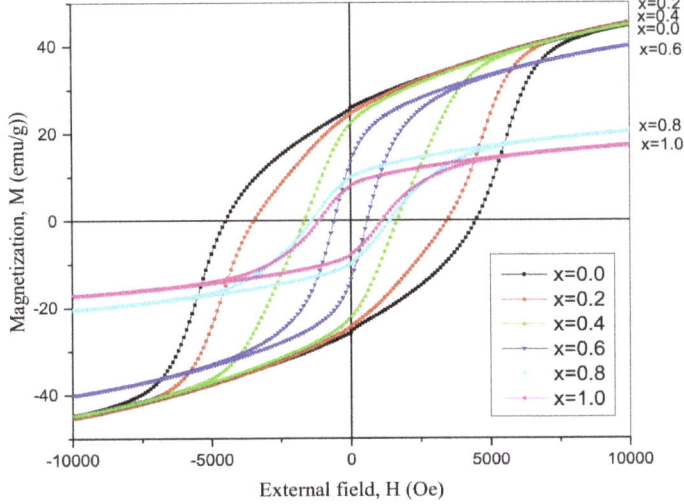

Figure 6. Hysteresis loops for the samples $BaFe_{12-x}Ti_xO_{19}$ ($x = 0.0 - 1.0$) sintered at 1000 °C.

Table 3 shows summary of the calculated values for the saturation magnetization, the remanence magnetization, the reduced magnetization, and the coercivity.

Table 3. The saturation magnetization, remanence magnetization, reduced remnant magnetization and coercive field for the system $BaFe_{12-x}Ti_xO_{19}$ ($x = 0.0$–1.0).

x	M_s (emu/g)	M_r (emu/g)	M_{rs} (emu/g)	H_c (kOe)
0.0	44.65	23.0606	0.5165	4.51
0.2	45.24	19.7474	0.4365	3.4557
0.4	44.83	19.5835	0.4368	1.6554
0.6	39.99	12.9342	0.3234	0.583
0.8	20.39	9.5829	0.47	1.385
1.0	17.17	7.7163	0.4494	1.11455

The saturation magnetization shows inverse relation titanium concentration. The partial substitution of Fe by titanium ions in these hexaferrites has affected the values of M_S, which depends on the amount of Ti ions diffusing into the BaM crystal in which Ti^{+3} ions partially replace Fe^{+3}. However, the higher the Ti concentration results in lower M_s because other phases are then formed as was concluded from XRD data.

Hard ferrites typically have a high value of M_r, which is about 50% of the saturation magnetization (M_s), therefore the reduced remnant magnetization ($M_{rs} = M_r/M_s$) must be nearly 0.5, and this is true for samples consisting of single-domains and random orientations of grains [32]. Table 3 shows that the values of M_{rs} are very close to 50%. The coercive field, H_c, for the samples varied with Ti concentration and it ranged between 0.6 and 4.5 kOe as can be seen in Table 3. A plot of M_s and H_c versus the concentration of titanium is shown in Figure 7.

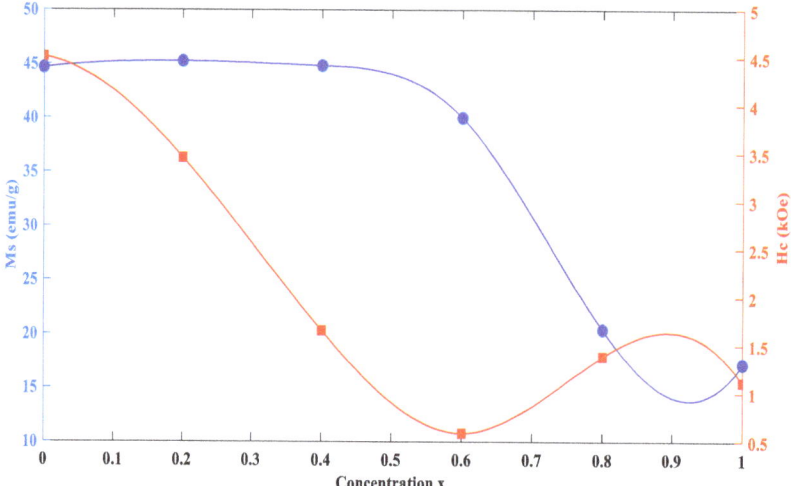

Figure 7. Saturation magnetization and coercive field versus Ti concentration.

We calculated the average paramagnetic behavior for each sample by approximating the loop by a straight line as shown in Figure 8. Of course, this behavior is correct only for shallow applied fields, but we found that there is a general trend in these samples that can be depicted by the slopes of these lines. Slopes of these lines are listed in Table 4. The slopes are related to the magnetic susceptibility and increased slowly by increasing the concentration up to $x = 0.6$ and then decreases to half its value after that.

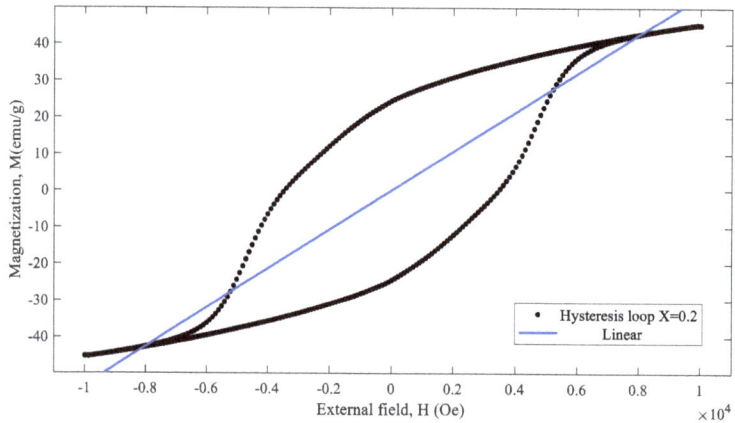

Figure 8. The hysteresis loop for the system $BaFe_{12-x}Ti_xO_{19}$ ($x = 0.2$) approximated by a straight line.

Table 4. The slope and hysteresis losses (in W) for the system $BaFe_{12-x}Ti_xO_{19}$ (x = 0.0–1.0).

Concentration (x)	Slope	Hysteresis Losses (W)
0.0	0.004908	518,580
0.2	0.005326	402,700
0.4	0.00571	212,180
0.6	0.005252	71,914
0.8	0.002623	84,982
1.0	0.002229	58,080

We also calculated hysteresis losses for each sample and the calculated values are listed in Table 4. It is clear from the values of losses that Ti concentration could be a scaling factor for magnetic energy stored when these materials are used in components of electromagnetic devices. This behavior is displaced in Figure 9. Losses are significant and almost linear for $x < 0.6$.

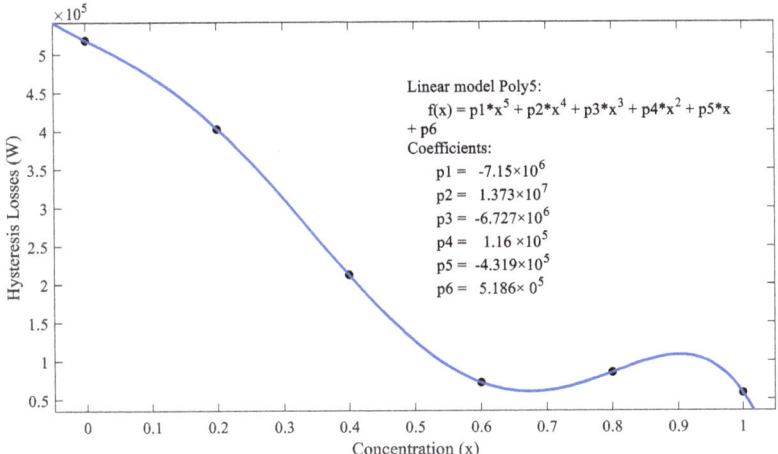

Figure 9. Hysteresis losses (in W) $BaFe_{12-x}Ti_xO_{19}$ (x = 0.0–1.0) sintered at 1000 °C.

4. Conclusions

A group of $BaFe_{12-x}Ti_xO_{19}$ samples with (x = 0.0–1.0) were prepared by sol-gel-auto combustion method and were sintered at 850 °C and 1000 °C for 1 h in air. The XRD data and IR data analysis confirmed the presence of M type phase of the hexaferrite in these samples, and the higher sintering temperature showed an improvement of crystallinity. The average crystallite size for all samples was in the range of 27–42 nm. We believe that these are suitable to obtain the reasonable signal to noise ratio when used in a high-density recording medium. The values of saturation magnetization (M_s) and retentivity (M_r) calculated values decreased with increasing the Ti concentration, and this may result from particle size effects. The addition of titanium reduced the saturation magnetization from about 45 emu/g down to 17.2 emu/g. The rest of the magnetic properties that include coercivity, remanence magnetization, reduced remnant magnetization, and hysteresis losses suggest that these materials are beneficial in the components of electromagnetic devices that are used in recording technology and permanent magnets. We suggest that further work is needed to study the effect of sintering temperatures and very low controlled Ti concentrations. In addition, structural and magnetic studies at temperatures above and below room temperature using other techniques could be of great importance.

Author Contributions: Formal analysis, L.A.A.-H.; Methodology, H.A.K.; Supervision, A.R.A.D. All authors contribute equally to this manuscript.

Funding: This research received no external funding.

Acknowledgments: This research work was supported by the Deanship of Graduate Studies and Research, Yarmouk University, Irbid, Jordan; project 7/2017. We acknowledge the help of Prof. A. Hammoudeh in preparation of samples and IR data.

Conflicts of Interest: The authors declare no conflict of interest.

References

1. Gutfleisch, O.; Willard, M.A.; Bruck, E.; Chen, C.H.; Sankar, S.G.; Liu, J.P. Magnetic materials and devices for the 21st century: Stronger, lighter, and more energy efficient. *Adv. Mater.* **2011**, *23*, 821–842. [CrossRef]
2. Cullity, B.D.; Graham, C.D. *Introduction to Magnetic Materials*; John Wiley & Sons: New York, NY, USA, 2011.
3. Bini, R.A.; Marques, R.F.C.; Santos, F.J.; Chaker, J.A.; Jafelicci, M., Jr. Synthesis and functionalization of magnetite nanoparticles with different amino-functional alkoxysilanes. *J. Magn. Magn. Mater.* **2012**, *324*, 534–539. [CrossRef]
4. Cochardt, A. Recent ferrite magnet developments. *J. Appl. Phys.* **1966**, *37*, 1112–1115. [CrossRef]
5. Sugimoto, M. The past, present, and future of ferrites. *J. Am. Ceram. Soc.* **1999**, *82*, 269–280. [CrossRef]
6. Pullar, R.C. Hexagonal ferrites: A review of the synthesis, properties and applications of hexaferrite ceramics. *Prog. Mater. Sci.* **2012**, *57*, 1191–1334. [CrossRef]
7. Koutzarova, T.; Kolev, S.; Ghelev, C.; Grigorov, K.; Nedkov, I. Structural and magnetic properties and preparation techniques of nanosized M-type hexaferrite powders. In *Advances in Nanoscale Magnetism*; Springer: Berlin, Germany, 2009; pp. 183–203.
8. Maswadeh, Y.; Mahmood, S.H.; Awadallah, A.; Aloqaily, A.N. Synthesis and structural characterization of nonstoichiometric barium hexaferrite materials with Fe:Ba ratio of 11.5–16.16. In *IOP Conference Series: Materials Science and Engineering*; IOP Publishing Ltd.: Philadelphia, PA, USA, 2015.
9. Radwan, M.; Rashad, M.M.; Hessien, M.M. Synthesis and characterization of barium hexaferrite nanoparticles. *J. Mater. Process. Technol.* **2007**, *181*, 106–109. [CrossRef]
10. Martirosyan, K.S.; Galstyan, E.; Hossain, S.M.; Wang, Y.-J.; Litvinov, D. Barium hexaferrite nanoparticles: synthesis and magnetic properties. *Mater. Sci. Eng. B* **2011**, *176*, 8–13. [CrossRef]
11. Meng, Y.Y.; He, M.H.; Zeng, Q.; Jiao, D.L.; Shukla, S.; Ramanujan, R.V.; Liu, Z.W. Synthesis of barium ferrite ultrafine powders by a sol–gel combustion method using glycine gels. *J. Alloy. Compd.* **2014**, *583*, 220–225. [CrossRef]
12. Kouřil, K. Local structure of hexagonal ferrites studied by NMR. (2013). Available online: https://www.researchgate.net (accessed on 20 March 2019).
13. Wu, M. M-Type barium hexagonal ferrite films. In *Advanced Magnetic MaterialsAdv*; InTech Open: London UK, 2012.
14. Awadallah, A.M.; Sami, M. Effects of preparation conditions and metal ion substitutions for barium and iron on the properties of M-type barium hexaferrites. Ph.D. Research Proposal, The University of Jordan, Amman, Jordan, 2012. [CrossRef]
15. Zhang, H.; Zeng, D.; Liu, Z. The law of approach to saturation in ferromagnets originating from the magnetocrystalline anisotropy. *J. Magn. Magn. Mater.* **2010**, *322*, 2375–2380. [CrossRef]
16. Bsoul, I.; Mahmood, S.H. Magnetic and structural properties of BaFe$_{12-x}$Ga$_x$O$_{19}$ nanoparticles. *J. Alloy. Compd.* **2010**, *489*, 110–114. [CrossRef]
17. Awawdeh, M.; Bsoul, I.; Mahmood, S.H. Magnetic properties and Mössbauer spectroscopy on Ga, Al, and Cr substituted hexaferrites. *J. Alloy. Compd.* **2014**, *585*, 465–473. [CrossRef]
18. Packiaraj, G.; Panchal, N.R.; Jotania, R.B. Structural and Dielectric studies of Cu substituted Barium hexaferrite prepared by Sol-gel auto combustion technique. *Solid State Phenom.* **2014**, *209*, 102–106. [CrossRef]
19. Mahmood, S.H.; Aloqaily, A.N.; Maswadeh, Y.; Awadallah, A.; Bsoul, I.; Juwhari, H. Structural and magnetic properties of Mo-Zn substituted (BaFe$_{12-4x}$Mo$_x$Zn$_{3x}$O$_{19}$) M-type hexaferrites. *Mater. Sci. Res. India* **2014**, *11*, 9–20. [CrossRef]
20. Haneda, K.; Hiroshi, K. Intrinsic coercivity of substituted BaFe$_{12}$O$_{19}$. *Japan. J. Appl. Phys.* **1973**, *12*, 355. [CrossRef]
21. Dushaq, G.H.; Mahmood, S.H.; Bsoul, I.; Juwhari, H.K.; Lahlouh, B.; AlDamen, M.A. Effects of molybdenum concentration and valence state on the structural and magnetic properties of BaFe$_{11.6}$Mo$_x$Zn$_{0.4-x}$O$_{19}$ hexaferrites. *Acta Metall. Sin. (English Letters)* **2013**, *26*, 509–516. [CrossRef]

22. Ali, I.; Islam, M.U.; Awan, M.S.; Ahmad, M. Effects of Ga–Cr substitution on structural and magnetic properties of hexaferrite (BaFe$_{12}$O$_{19}$) synthesized by sol–gel auto-combustion route. *J. Alloy. Compd.* **2013**, *547*, 118–125. [CrossRef]
23. Wang, S.; Ding, J.; Shi, Y.; Chen, Y.J. High coercivity in mechanically alloyed BaFe$_{10}$Al$_2$O$_{19}$. *J. Magn. Magn. Mater.* **2000**, *219*, 206–212. [CrossRef]
24. Ali, I.; Islam, M.U.; Awan, M.S.; Ahmad, M.; Ashiq, M.N.; Naseem, S. Effect of Tb^{3+} substitution on the structural and magnetic properties of M-type hexaferrites synthesized by sol–gel auto-combustion technique. *J. Alloy. Compd.* **2013**, *550*, 564–572. [CrossRef]
25. Sharma, R.; Bisen, D.P.; Shukla, U.; Sharma, B.G. X-ray diffraction: a powerful method of characterizing nanomaterials. *Recent Res. Sci. Technol.* **2012**, *4*, 77–79.
26. Uvarov, V.; Popov, I. Metrological characterization of X-ray diffraction methods at different acquisition geometries for determination of crystallite size in nano-scale materials. *Mater. Charact.* **2013**, *85*, 111–123. [CrossRef]
27. Stuart, B.H. Experimental methods. In *Infrared Spectroscopy: Fundamentals And Applications*; John Wiley & Sons, Ltd.: Hoboken, NJ, USA, 2005; pp. 15–44.
28. Coates, J. Interpretation of infrared spectra, a practical approach. In *Encyclopedia of Analytical Chemistry*; John Wiley & Sons Ltd.: Chichester, UK, 2000; pp. 10815–10837.
29. Goldstein, J.; Yakowitz, H. *Practical Scanning Electron Microscopy: Electron and Ion Microprobe Analysis*; Springer: Boston, MA, USA, 1975.
30. Aravind, A. Synthesis and characterization of 3d-transition metals doped ZnO thin films and nanostructures for possible spintronic applications. Ph.D. Thesis, Cochin University of Science and Technology, Kerala, India, 2012.
31. Monshi, A.; Foroughi, M.R.; Monshi, M.R. Modified Scherrer equation to estimate more accurately nano-crystallite size using XRD. *World J. Nano Sci. Eng.* **2012**, *2*, 154–160. [CrossRef]
32. Mahmood, S.H.; Dushaq, G.H.; Bsoul, I.; Awawdeh, M.A.; Juwhari, H.K.; Lahlouh, B.; AlDamen, M.A. Magnetic properties and hyperfine interactions in M-type BaFe$_{12-2x}$Mo$_x$Zn$_x$O$_{19}$ hexaferrites. *J. Appl. Math. Phys.* **2014**, *2*, 77–87. [CrossRef]
33. Li, Y.; Wang, Q.; Yang, H. Synthesis, characterization, and magnetic properties of nanocrystalline BaFe$_{12}$O$_{19}$ Ferrite. *Curr. Appl. Phys.* **2009**, *9*, 1375–1380. [CrossRef]
34. Trukhanov, S.V.; Trukhanov, A.V.; Kostishyn, V.G.; Panina, L.V.; Trukhanov, A.V.; Turchenko, V.A.; Tishkevich, D.I.; Trukhanova, E.L.; Yakovenko, O.S.; Matzui, L.Y.; et al. Effect of gallium doping on electromagnetic properties of barium hexaferrite. *J. Phys. Chem. Solid.* **2017**, *111*, 142–152. [CrossRef]
35. Trukhanov, S.V.; Trukhanov, A.V.; Kostishyn, V.G.; Panina, L.V.; Trukhanov, An.V.; Turchenko, V.A.; Tishkevich, D.I.; Trukhanova, E.L.; Yakovenko, O.S.; Matzui, L.Yu. Investigation into the structural features and microwave absorption of doped barium hexaferrites. *Dalton Trans.* **2017**, *46*, 9010–9021. [CrossRef]
36. Trukhanov, S.V.; Trukhanov, A.V.; Salem, M.M.; Trukhanova, E.L.; Panina, L.V.; Kostishyn, V.G.; Darwish, M.A.; Trukhanov, A.V.; Zubar, T.I.; Tishkevich, D.I.; et al. Preparation and investigation of structure, magnetic and dielectric properties of (BaFe$_{11.9}$Al$_{0.1}$O$_{19}$)$_{1-x}$-(BaTiO$_3$)$_x$ bicomponent ceramics. *Ceram. Int.* **2018**, *44*, 21295–21302. [CrossRef]
37. Trukhanov, S.V.; Trukhanov, A.V.; Turchenko, V.A.; Trukhanov, An.V.; Tishkevich, D.I.; Trukhanova, E.L.; Zubar, T.I.; Karpinsky, D.V.; Kostishyn, V.G.; Panina, L.V.; et al. Magnetic and dipole moments in indium doped barium hexaferrites. *J. Magn. Magn. Mater.* **2018**, *457*, 83–96. [CrossRef]
38. Trukhanov, S.V.; Trukhanov, A.V.; Turchenko, V.A.; Kostishyn, V.G.; Panina, L.V.; Kazakevich, I.S.; Balagurov, A.M. Structure and magnetic properties of BaFe$_{11.9}$In$_{0.1}$O$_{19}$ hexaferrite in a wide temperature range. *J. Alloy. Compd.* **2016**, *689*, 383–393. [CrossRef]
39. Turchenko, V.A.; Trukhanov, S.V.; Balagurov, A.M.; Kostishyn, V.G.; Trukhanov, A.V.; Panina, L.V.; Trukhanova, E.L. Features of crystal structure and dual ferroic properties of BaFe$_{12-x}$Me$_x$O$_{19}$ (Me = In^{3+} and Ga^{3+}; x = 0.1–1.2). *J. Magn. Magn. Mater.* **2018**, *464*, 139–147. [CrossRef]

© 2019 by the authors. Licensee MDPI, Basel, Switzerland. This article is an open access article distributed under the terms and conditions of the Creative Commons Attribution (CC BY) license (http://creativecommons.org/licenses/by/4.0/).

Article

Design and Analysis of a Plate Type Electrodynamic Suspension Structure for Ground High Speed Systems

Zhaoyu Guo [1], Danfeng Zhou [1,2], Qiang Chen [1], Peichang Yu [1] and Jie Li [1,*]

[1] Maglev Research Center, College of Intelligence Science and Technology, National University of Defense Technology, Changsha 410073, China
[2] State Key Laboratory of Functional Materials for Informatics, Shanghai Institute of Microsystem and Information Technology, Chinese Academy of Sciences, Shanghai 200050, China
* Correspondence: jieli@nudt.edu.cn

Received: 8 July 2019; Accepted: 29 August 2019; Published: 4 September 2019

Abstract: The research of ground high speed systems has been popular, especially after the announcement of Hyperloop concept, and the analysis of the suspension structure is critical for the design of the system. This paper focuses on the design and analysis of a plate type electrodynamic suspension (EDS) structure for the ground high speed system. The working principle of proposed whole system with functions of levitation, guidance and propulsion is presented, and the researched EDS structure is composed of permanent magnets (or superconducting magnets) and non-ferromagnetic conductive plates. Levitation and guidance are achieved by forces generated through the motion of the magnets along the plates. The plate type EDS structure is analyzed by three-dimensional (3D) finite element method (FEM) in ANSYS Maxwell. Structure parameters that affect the EDS performances are investigated, which include dimensions of magnets and plates, plate material, the relative position between magnets and plates, and arrangement of magnets. The properties of forces are discussed, especially for the levitation force, and the levitation working point is decided based on the analysis. Levitation-drag ratio of the plate type structure is investigated, and it improves with the increasing of vehicle velocity. The analysis results indicate that the plate type EDS structure is feasible for applications in ground high speed systems. The following study will focus on the dynamic research of the EDS system.

Keywords: magnetic levitation; electrodynamic structure; ground high speed system; finite element analysis

1. Introduction

Nowadays, the research of ground high speed systems including high speed train and super speed test sled has been popular with the increasing demand for higher speed. For example, Elon Musk presented the concept of Hyperloop in 2013 and the Evacuated Tube Transportation (ETT) has received much attention from research institutes and the transportation industry [1,2]. Compared with wheel-rail structure, contactless design is expected to be adopted in ground high speed systems to minimize frictions, vibrations and noises during the operation. Magnetic levitation (maglev) is the most widely studied contactless technology and the maglev train is a well-known application. The maglev train has a different working principle from conventional wheel-rail train driving forward by frictions, as it achieves levitation through interaction between the magnets aboard and the rails and produces propulsion force electromechanically without contact with the rails [3].

The Electrodynamic Suspension (EDS) repulsion system and the Electromagnetic Suspension (EMS) attraction systems are two main kinds of maglev systems [4]. The EDS system achieves levitation by the electromagnetic repulsive force between the magnet and the non-ferromagnetic conductive rail. The relative motion of the magnet along the rail will induce eddy current in the ground conductor, and according to Lenz law, the magnetic field induced by the eddy current will oppose the magnetic field from the aboard magnet to generate repulsive force [5]. When the moving vehicle reaches a certain speed, the repulsive force increases enough to levitate the train. Thus, the EDS structure does not need active control and is an essential open-loop stable system [6]. Superconducting (SC) magnet and permanent magnet (PM) are two excitation sources applied in EDS system, and there are two kinds of EDS rails: continuous conductive plate and discrete metal coil. The Holloman High Speed Test Track (after maglev update) is a typical continuous EDS structure with copper plates embedded in the guideway, and the SC magnets aboard interact with the plates to provide levitation and guidance forces through relative motion [7]. The sled is propelled by solid fuel rocket motors, and the test speed reached 673 km/h in 2008. However, the rocket motor is expensive and cannot be reused, and it is hard to adjust the running speed during operation. The publications about Holloman High Speed Test Track are few and about the program update and some flight test results without critical characteristics analysis mentioned. A representative of typical discrete EDS system is the SC maglev train of JR Company in Japan, which adopts figure-eight null flux coils to achieve a high levitation-drag ratio [8]. The levitation coil and the propulsion coil are at the same side of the SC magnet, and the ground transportation speed record of 603 km/h was made by the SC maglev train on the Yamanashi Line in 2015. Unlike the EDS system, electromagnet is adopted in EMS structure to interact with the ferromagnetic rail to generate attractive electromagnetic force. EMS is an open-loop unstable system and active control of currents in the electromagnet is needed to achieve levitation at the expected gap [9,10]. Now, the top operation speed of EMS train is 430 km/h from the TR train on Shanghai maglev line [11].

EMS and EDS both have advantages in the application. Through active control of currents in the electromagnet, accurate levitation gap could be realized in EMS system and the dynamic performance during operation is adjustable. However, the electromagnet is heavy and the ancillary equipment, such as power supply, signal detection, and control systems, make the whole structure complex. Additionally the operation of high speed will make the conduction of real-time control difficult. Another problem is the obvious eddy current in the steel rail at high speeds [12,13], which will reduce the levitation force and lead to increasing currents and heat in the electromagnet. Compared with EMS, EDS has been attracting more attention in the research of ground high speed systems. The design and analysis of a Hyperloop structure including levitation and propulsion control system was conducted by Abdelrahman et al. [14], and null flux coil structure was adopted. An all-in-one system containing functions of levitation, propulsion and guidance was presented by Ji et al., and conductive plate EDS structure was adopted in the design [15]. A high-temperature superconducting EDS system used in the high speed maglev train was studied by Hao et al.; a mirror method was proposed for calculating and the accuracy was verified by the FEM model [16].

This paper focuses on the design and analysis of a plate type EDS structure for ground high speed systems. Since fluctuation of electrodynamic forces in coil type EDS structure is obvious due to discrete arrangement of coils [17] and the manufacture of coils is relatively complex, conductive plate rails are adopted to interact with aboard moving magnets to provide levitation and guidance. Meanwhile, magnetic fields from the other side of the magnets are used for propulsion, which could simplify the whole system. The EDS structure is analyzed by finite element method, and parameters of magnets and plates that affect the EDS performance are investigated. The forces properties at constant speeds are analyzed and the analysis results have shown that the researched plate type EDS structure is promising for application in the ground high speed systems.

2. Proposed Ground High Speed System

The designed ground high speed structure is composed of two subsystems, levitation system and propulsion system, as shown in Figure 1. Axis X is the lateral direction, axis Y is the vehicle moving direction, and axis Z is the vertical direction where the vehicle is levitated. The vehicle body is mounted upon the propulsion stator and magnets are placed on both lateral sides of the vehicle. SC magnet and PM can be adopted as excitation source in the design, and PM is selected in this paper. Non-ferromagnetic metal plates (copper plates or aluminum plates) are embedded in the lateral symmetrical guideway beams, and the supporting structures including beams and buttress are made of non-metallic material to avoid effect on propulsion and levitation. When the magnet aboard moves along the rails, eddy current is induced in the plates and the interaction between the magnet and the eddy current will generate levitation, guidance and drag forces. The double plate (upper and lower) structure is adopted and the center-line of the magnets is lower than that of the double plates in axis Z. The overlap area between the magnet and the lower plate is larger than the one with the upper plate, and the resultant on the magnet in axis Z will be upward levitation force. The linear synchronous motor contains the magnet mover and the long hollow stator coils. It can be seen that both sides of magnets are used in the system to get high utilization of magnetic fields and simplify the whole structure. Traveling wave magnetic fields are generated by flowing three-phase AC in the hollow stator coils, and the magnetic fields interact with the magnet aboard to achieve electromagnetic thrust and synchronous motion. The deceleration is conducted by reversing the direction of the stator currents to produce braking forces, and the drag forces from metal plates could also make a part of contribution.

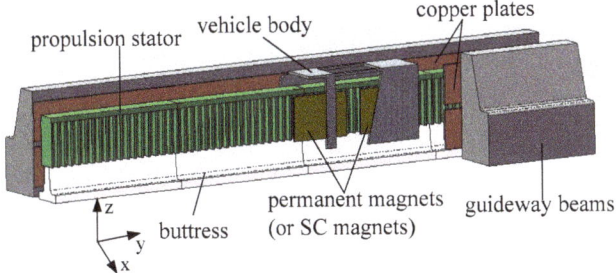

Figure 1. Proposed structure of ground high speed system.

3. Analysis of the Plate Type EDS Structure

The double plate structure could be divided to two single plates, and the magnetic forces acting on the magnet are the resultant of component forces from single plates. It is reasonable to analyze properties of the single plate structure for simplicity, and the side view of the single plate structure is shown in Figure 2.

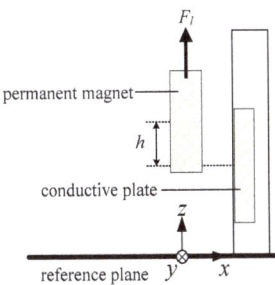

Figure 2. Single plate electrodynamic suspension (EDS) structure.

According to the Ampere circuital theorem, the relationship between the induced current density J and the magnetic flux density B in the plate can be described as:

$$\begin{vmatrix} \vec{i} & \vec{j} & \vec{k} \\ \frac{\partial}{\partial x} & \frac{\partial}{\partial y} & \frac{\partial}{\partial z} \\ B_x & B_y & B_z \end{vmatrix} = \mu(J_x + J_y + J_z) \tag{1}$$

Then:

$$\begin{cases} J_x = \frac{1}{\mu}(\frac{\partial B_z}{\partial y} - \frac{\partial B_y}{\partial z}) \\ J_y = \frac{1}{\mu}(\frac{\partial B_x}{\partial z} - \frac{\partial B_z}{\partial x}) \\ J_z = \frac{1}{\mu}(\frac{\partial B_y}{\partial x} - \frac{\partial B_x}{\partial y}) \end{cases} \tag{2}$$

Based on Lorentz law, the levitation, drag and guidance forces between the plate and the magnet are:

$$\begin{array}{l} F_l = \iiint (J_x B_y - J_y B_x) dV \\ F_d = \iiint (J_x B_z - J_z B_x) dV \\ F_g = \iiint (J_y B_z - J_z B_y) dV \end{array} \tag{3}$$

where V is the integral domain in the plate and μ is the permeability of the plate. The magnetic field and force equations are complex to solve, and approximations are usually involved in the derivation [18–20]. The analytical approach has its limited applicability in the initial design of the plate type EDS structure, and finite element simulation (FEM) using ANSYS Maxwell is adopted to study the characteristics of the EDS magnetic forces. The parameters of proposed EDS structure for FEM simulation are listed in Table 1, which are adopted in the construction of a prototype platform for test. The mesh plot of the simulation model in ANSYS and the pole pith of the two magnets are shown in Figure 3, and the following analysis is on the constant speed in axis Y without electrodynamic terms in axes X and Z.

Table 1. Parameters of the FEM model. PM: permanent magnet.

Variable	Symbol	Value	Unit
Length of magnet (Y axis)	lm	230	mm
Height of magnet (Z axis)	Hm	200	mm
Thickness of magnet (X axis)	Tm	20	mm
Pole pitch of magnets (Y axis)	τ	270	mm
Remanence of PM	B_r	1.3	T
Coercivity of PM	H_c	940	kA/m
Relative permeability of PM	μ_r	1.09	
Height of plate (Z axis)	Hp	200	mm
Thickness of plate (X axis)	Tp	8	mm
Gap between magnet and plate (X axis)	G	10	mm
Height difference between mid-lines of magnet and plate (Z axis)	h	70	mm

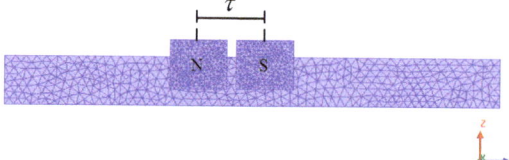

Figure 3. Mesh plot of the simulation model in ANSYS.

3.1. Levitation Working Point

The height difference h is an important parameter in the EDS structure, and it can be regarded as the working point of levitation. The simulation is conducted at the speed of 10 m/s, and the results of three forces are shown in Figure 4. The guidance and the drag forces have the same trend. The values decrease with the increasing of the height difference h, and this can be explained by the change of overlap area between the magnet and the plate in qualitative analysis. However, the levitation force is different from the other two. At small values of h, the levitation force increases with the growing of the height difference h. It reaches the top value at around 60 mm, and then stays relatively constant for a range of values. Then, with the further increase of h, the levitation force will decrease as the other two forces. The trend of levitation force could also be explained qualitatively. When $h = 0$, the magnet and the plate are mid-lines coincided, the levitation force is 0 due to symmetry. With the increasing of h, the symmetry is destroyed and the levitation force appears and increases. Then, with the further increasing of h at a large value, the interaction between the magnet and the plate will be very weak and the forces will all be zero in the end.

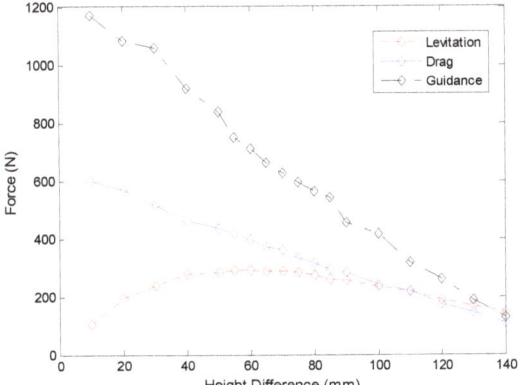

Figure 4. EDS forces as a function of height difference h (speed 10 m/s).

The levitation-drag ratio is shown in Figure 5, and it grows with the increase of h. The height difference h is set to 70 mm (about 1/3 of magnet height 200 mm) to provide adequate levitation force, as listed in Table 1. However, the levitation force around the working point changes little, and the system could not provide sufficient levitation stiffness at current setting. Because the PM is unable to generate strong enough magnetic fields, the working point cannot be set in the domain with large height differences (such as around 110 mm in Figure 4). Thus, the height difference h can be set around point of 1/3 height of magnet to provide enough levitation force when PM is used, and the working point can be moved to a larger h beyond 1/3 height of magnet to achieve stronger levitation stiffness if SC magnet is adopted in the system.

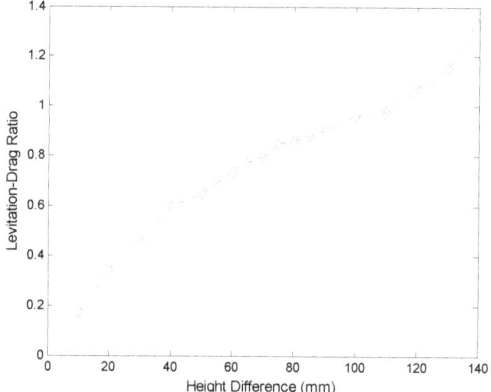

Figure 5. Levitation-drag ratio as a function of height difference h at a speed of 10 m/s.

3.2. Guidance Stiffness

Guidance stiffness is an important performance index in the maglev system, which is related with lateral stability. The guidance stiffness of the system can be studied by changing the gap G between the magnet and the plate at speeds of 10 m/s and 100 m/s, respectively. The guidance force results are shown in Figure 6. The value of G is set to 10 mm in the system as shown in Table 1. When the gap from one side of the vehicle body moves to 8 mm, then the gap value on the other side will be 12 mm. The guidance force on the vehicle from both sides of plates will be $\Delta F = 683\text{ N} - 590\text{ N} = 93\text{ N}$, and the guidance stiffness 23 N/mm is calculated at the speed of 10 m/s. Similarly the guidance force and the stiffness at the speed of 100 m/s are 249 N and 62 N/mm, respectively. The stiffness at low speeds is relatively insufficient compared with that at a high speed, and guidance wheels could be designed in the EDS structure.

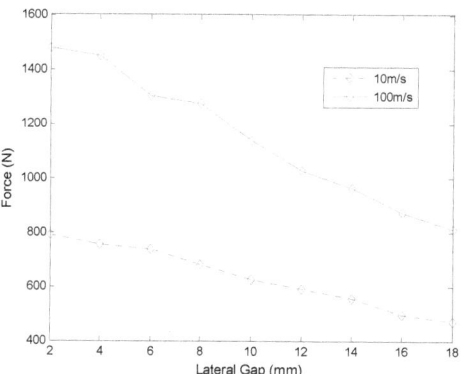

Figure 6. Guidance force as a function of the gap between the magnet and the plate.

3.3. Effects of Speed

Since the EDS force is based on the relative motion between the magnet and the plate, it is essential to research the effects of vehicle speed on magnetic forces. The forces in three axes at different velocities with the structure parameters in Table 1 are shown in Figure 7. The performances of copper and aluminum plates are compared in the simulation.

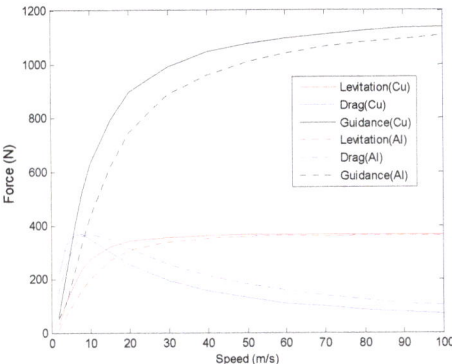

Figure 7. EDS forces as a function of magnet velocity.

The results show that the levitation force and the guidance force increase with magnet velocity at initial low speeds; then, they will stay relatively constant when reaching high speeds. This is beneficial for the super speed system that the system can be steadily levitated on the set height at speeds beyond the critical point, and the constant guidance force will not heighten the demand on structural strength of the vehicle body. The performance of drag force is different from the other two forces. It increases at the initial low speeds and after reaching the peak value, the force decreases with the increasing of magnet speed. Thus, the ratio of the levitation force and the drag force becomes higher at relative high velocities, as shown in Figure 8, which is beneficial for application in the ground high speed system. Additionally, it can be seen that the EDS structure with copper plate could provide better levitation-drag ratio than aluminum plate.

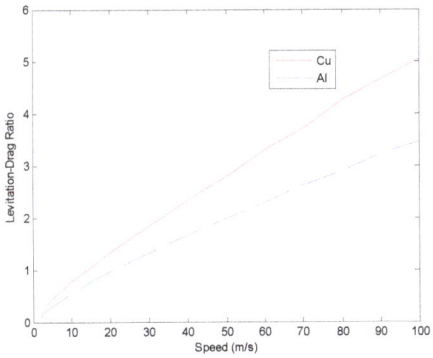

Figure 8. Levitation drag ratio over magnet velocity.

Magnetic forces are generated by reactions between the magnets and the eddy currents in the plates. It is essential to study the eddy current property, and the eddy current distributions on the plate surface at different vehicle speeds are shown in Figure 9. To make the analysis easier to understand, only one permanent magnet is adopted here and the principles are same for NS poles arranged magnets. In the simulation models, the magnets move from right to left and the rainbow color presents the magnitude of induced eddy current densities. It can be seen that the maximum value of eddy current on the plate becomes larger with the increasing of speed at first (for example from 2 m/s to 10 m/s) and then will stay almost constant at high speeds (for example at 70 m/s and 100 m/s). The simplified illustration of interaction between the magnet and the plate with single magnet in cross-section view is shown in Figure 10.

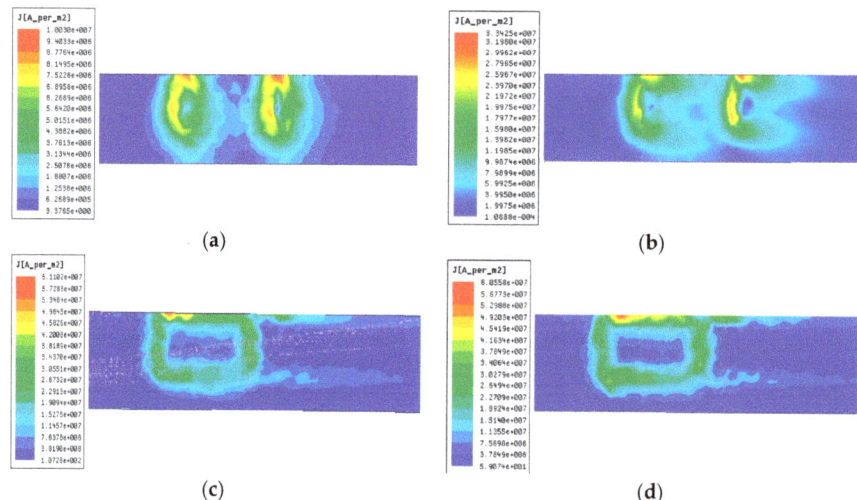

Figure 9. Distribution of eddy currents on plate surface at different speeds with single magnet, (**a**) 2 m/s, (**b**) 10 m/s, (**c**) 70 m/s, and (**d**) 100 m/s.

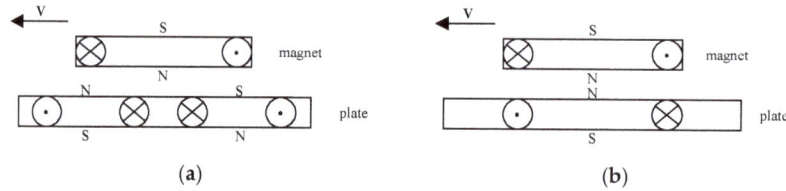

Figure 10. Simplified illustration of interaction between the magnet and the plate at different speeds with single magnet in cross-section view: (**a**) low speed and (**b**) high speed.

At low speeds, there exist two symmetric vortices of eddy current distributions. From the Lenz law, the current flow directions are opposite in the two vortices as illustrated in Figure 10a and the interactions between both vortices and the magnet perform as drag force in the moving direction. With the increasing of magnet speed, the back vortex will be pushed and reduced. Thus, there will be only one vortex with perfect same shape as the magnet as shown in Figure 9c,d at high speeds. It can be seen from Figure 10b that the drag force will be greatly reduced and the levitation-drag ratio will improve obviously.

3.4. Magnet Design

3.4.1. Distribution of Magnetic Field

PM and SC magnet are two main magnets adopted in the EDS system. Although permanent magnet is adopted in this analysis, it is important to study the magnetic field distributions of both magnets. The parameters of both magnets for comparison are listed in Table 2, and they have the same shape and size.

The rainbows of magnetic field vector from both magnets are shown in Figure 11, and the amplitudes of magnetic flux density at the mid-lines in Y and Z directions 10 mm away from the surface of magnets are plotted in Figure 12. It can be seen that the distribution tendencies of magnetic fields from PM and SC magnet are basically same. The magnetic fields near the borders of magnets are strongest and will decrease apart from the borders. The same distributions could indicate that the analysis results of PM are also appropriate for SC magnet, and the difference is that the SC could provide much stronger magnetic fields through adjusting the flowing currents in the SC coils.

Table 2. Parameters of both magnets. SC: superconducting.

Variable	Value	Unit
Length	230	mm
Height	200	mm
Thickness	20	mm
Remanence of PM	1.3	T
Coercivity of PM	940	kA/m
Relative permeability of PM	1.09	
Current in SC magnet	18.8	kA

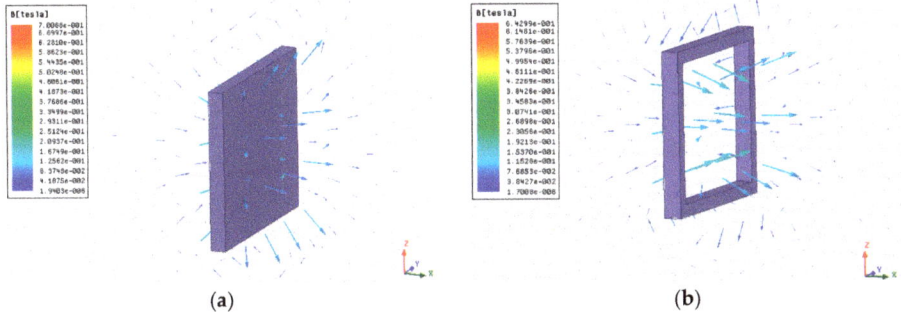

Figure 11. Rainbows of magnetic field vector, (a) PM, (b) SC magnet.

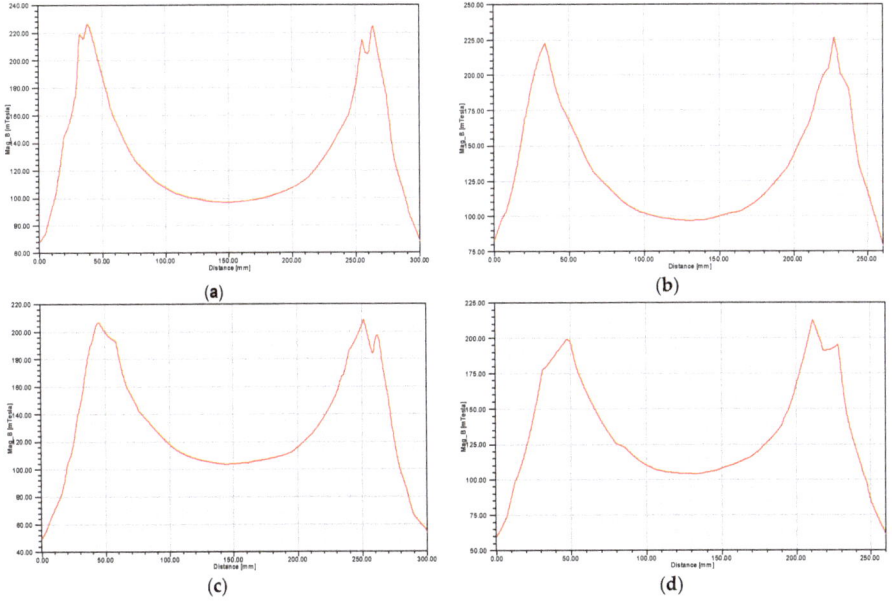

Figure 12. Distributions of magnetic flux density: (a) Y direction of PM, (b) Z direction of PM, (c) Y direction of SC magnet, and (d) Z direction of SC magnet.

3.4.2. Magnet Thickness

The height and the pole length of the permanent magnets are decided by the design of the propulsion, which is not involved here. The magnet pole face is 23 cm × 20 cm, the remanence of the NdFeB magnet is 1.3 T and the coercivity is 940 kA/m. Magnetic flux densities of the center point on the magnet surface are illustrated in Figure 13. As can be seen, although the magnetic flux density strengthens with increasing magnet thickness, the increment becomes gentle with the increasing of the thickness. Considering the manufacture and the magnet weight, it is not a good idea to get strong magnetic field through increasing the magnet thickness and 20 mm is selected as listed in Table 1.

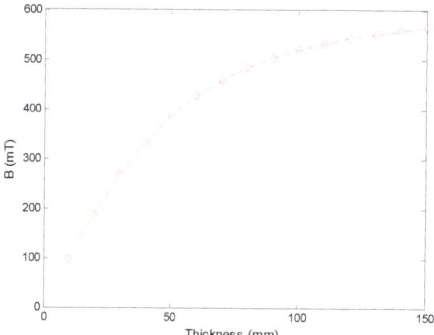

Figure 13. Magnetic flux density over magnet thickness.

3.4.3. Magnet Length

The effects of different magnet lengths on levitation force are researched. The pole length of the permanent magnets is 270 mm, which is decided from the propulsion calculation. The results are shown in Table 3, and the speed is 10 m/s. Although levitation force becomes larger with the increase of magnet length, the increment is narrow when the magnet length approaches the pole length. Furthermore, the long and large permanent magnet is fragile and difficult to manufacture, thus a 230 mm long magnet was selected in the design.

Table 3. Levitation forces with different magnet lengths.

Magnet Length (mm)	Levitation Force (N)
190	230
200	236
210	240
220	256
230	268
240	280
250	288
260	295
270	297

3.4.4. Magnet Arrangement

Although the magnets are alternately arranged in magnetic N and S poles for propulsion in the proposed system as shown in Figure 3, it is still helpful to research the influence of magnet arrangement on magnetic forces, especially the levitation force. Three kinds of magnet arrangements are studied: single magnet, same pole arranged magnets (such as NN), and NS poles arranged magnets (shown in Figure 3); the results are shown in Figure 14. Double value of single magnet arrangement is included to be as a reference.

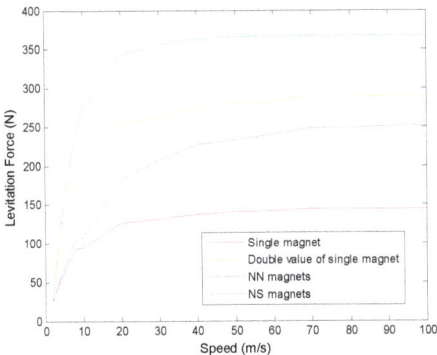

Figure 14. Levitation forces with different magnet arrangements (NN and NS) over magnet velocity.

It can be seen that the levitation force value of NS arranged magnets is larger than the double value of single magnet, and the value of NN arranged magnets is smaller than the double one. This can be explained in qualitative analysis. For the NN arranged magnets, the change rate of magnetic field generated by the back N pole magnet in conductive plate is gentler than the one by the front magnet. According to Lenz law, the back magnet will generate weaker induced voltage and eddy current, which will make the levitation force of NN arranged magnets smaller than the double value of single magnet. However, the change rate of magnetic field generated by the back magnet in conductive plate is more dramatic than the one by the front magnet. Then, the back magnet will generate stronger induced voltage and eddy current, which will make the levitation force of NS arranged magnets larger than the double value of single magnet. The eddy current distributions of NN and NS arranged magnets at speeds of 10 m/s and 100 m/s are shown in Figure 15. It is obvious that the NS arranged magnets could induce stronger eddy currents than NN arranged magnets, which has verified the explanation. From the above analysis, we can also get that there exist difference between the two forces from the front and the back magnets, and this will be studied in the following.

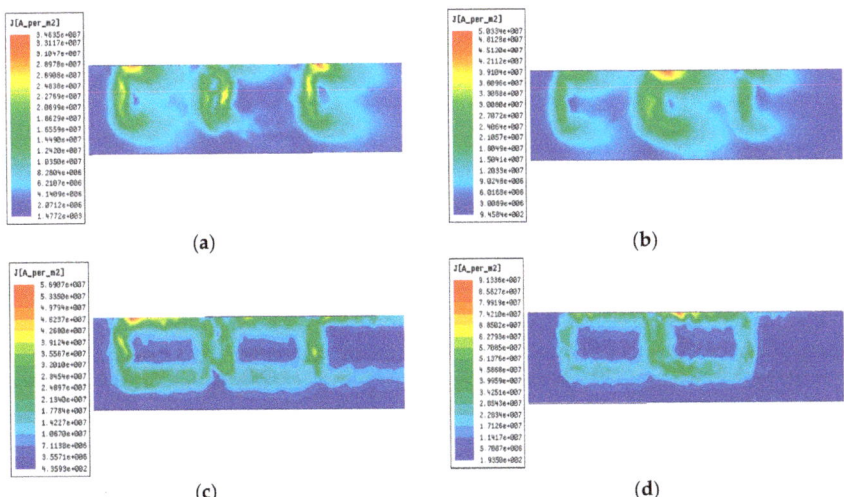

Figure 15. Distribution of eddy currents on plate surface at different speeds with NN and NS arranged magnets: (**a**) NN magnets at 10 m/s, (**b**) NS magnets at 10 m/s, (**c**) NN magnets at 100 m/s, and (**d**) NS magnets at 100 m/s.

3.5. Plate Design

3.5.1. Plate Materials

In the proposed EDS structure, the plate should be non-ferromagnetic conductive. Two common materials copper and aluminum are separately studied, whose conductivities are 58 MS/m and 38 MS/m, respectively. The comparisons of forces and levitation-drag ratios are shown in Figures 7 and 8, respectively. The levitation force and the guidance force from the copper plate increase faster than that from the aluminum plate at low speeds, but the difference becomes minor when the magnet speed is high and the forces tend to be the same. The drag force from the copper plate is larger than the aluminum one at the initial low speeds, and after the peak value it decreases faster than the aluminum one. Levitation-drag ratio is an important index, and from Figure 8, the EDS structure with copper plate could provide a higher levitation-drag ratio than aluminum plate.

3.5.2. Plate Height

Plate height is an important factor that will influence the interaction area between the magnet and the plate. The simulation results of different plate heights with constant height difference h 70 mm at speed of 10 m/s are shown in Figure 16. The forces in three directions including levitation drag and guidance grow with the increasing of plate height, and the guidance force increases fastest. Considering high plate will bring large drag force and high construction cost, 200 mm (the same height with PM) is adopted in the system design.

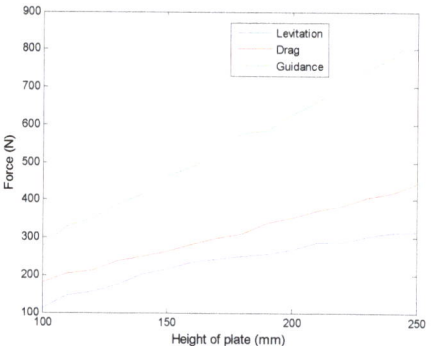

Figure 16. Levitation force as a function of plate height.

3.5.3. Plate Thickness

Plate thickness is another important parameter to be designed in the plate type EDS structure. Because of the skin effect at high speeds, the analysis is selected at speed of 100 m/s. The three forces and levitation-drag ratio are shown in Figure 17. It can be seen that the levitation and the guidance forces become larger with the increasing of thickness at first and will stay constant, and the drag force decreases with the increasing of thickness and the change becomes mild at relatively large thickness. Thus, the levitation-drag ration improves with the increasing of plate thickness. Considering the construction cost, 8 mm is selected in the system design.

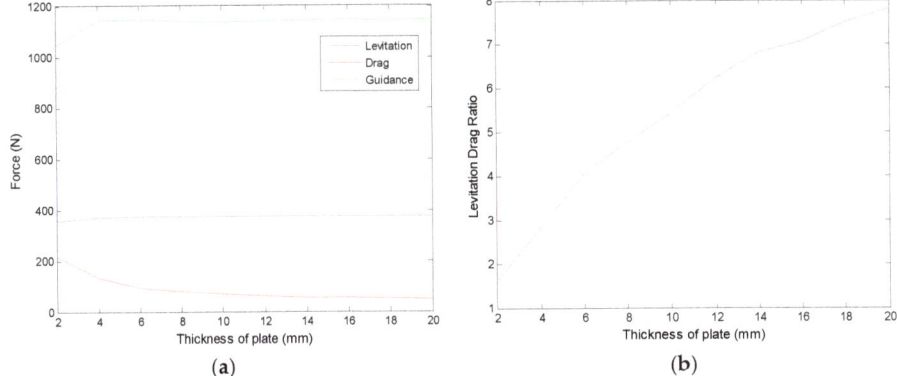

Figure 17. Simulation results of different thicknesses of plate at speed of 100 m/s. (**a**) Levitation, drag, and guidance forces, (**b**) levitation drag ratio.

3.6. Distribution of Levitation Forces on NS Magnets

In the above analysis in Section 3.4.4, it is explained that there is difference between the levitation forces of the front magnet and the back magnet in the NS arranged magnets. This is important to research, since the uneven distribution of levitation forces will lead to imbalance of vehicle during operation. The comparison of levitation forces of the front and the back magnets at different speeds are shown in Figure 18. The levitation force from the back magnet increases faster than that from the front magnet at the low speeds, and then it decreases slightly until the two forces are same value. The non-uniform distribution of levitation forces on NS magnets during the vehicle operation is obvious when the vehicle is at a low or middle speed, which indicates a higher standard for the vehicle design.

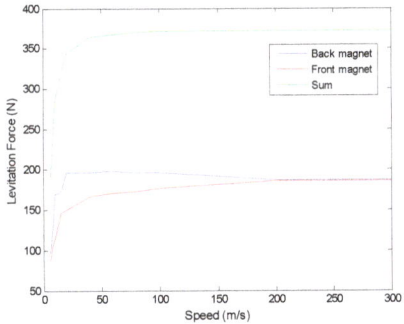

Figure 18. Simulation results of levitation forces of the front and the back magnets.

4. Conclusions

A plate type EDS structure proposed for ground high speed systems is analyzed in this paper. The ground high speed system is integrated with functions of levitation, propulsion and guidance, and the EDS structure is composed of non-ferromagnetic conductive plate and PM (or SC magnet). The designed plate type EDS structure is researched by FEM simulations, and characteristics of the EDS structure are analyzed based on the results.

- The levitation working point is better to be set around 1/3 height of PM to get enough levitation force, and it can be moved to a larger h when SC magnet is applied in the design.
- The lateral stiffness at low speeds is relatively insufficient compared with that at high speeds, and guidance wheels can be adopted.

- The effects of different speeds on magnetic forces are studied and the levitation-drag ratio improves with the increasing of magnets speed.
- PM and SC magnet have the same distribution of magnetic field, and the proper thickness and length of magnet are set. NS arranged magnets show better performance than NN arranged magnets.
- Copper plate could provide larger levitation force and higher levitation-drag ratio than aluminum plate. Proper height and thickness of plate are decided based on the performance and the construction cost.
- The non-uniform distribution of levitation forces on NS magnets will disappear at high speeds.

The results show that the proposed EDS structure is a promising option for application in ground high speed systems. Future research will focus on the study of dynamics properties of the plate type EDS structure, especially the under-damped characteristics of EDS structure during operation.

Author Contributions: Conceptualization, Z.G. and J.L.; methodology, D.Z.; software, Z.G.; formal analysis, Q.C.; investigation, P.Y.; writing—original draft preparation, Z.G.; writing—review and editing, D.Z.; supervision, J.L.; project administration, J.L.

Funding: This research was supported by the National Key R&D Program of China (No. 2016YFB1200601), and the Opening Foundation of the State Key Laboratory of Functional Materials for Informatics (No. SKL-2017-07).

Conflicts of Interest: The authors declare no conflict of interest.

References

1. Palacin, R. Hyperloop, the electrification of mobility, and the future of rail travel. *IEEE Electr. Mag.* **2016**, *4*, 4–51. [CrossRef]
2. Hyperloop Alpha. Available online: http://www.spacex.com/hyperloopalpha (accessed on 12 August 2013).
3. Lee, H.W.; Kim, K.C.; Lee, J. Review of maglev train technologies. *IEEE Trans. Magn.* **2006**, *42*, 1917–1925. [CrossRef]
4. Yan, L. Development and application of the maglev transportation system. *IEEE Trans. Appl. Supercond.* **2008**, *18*, 92–99. [CrossRef]
5. Davey, K. Analysis of an electrodynamic Maglev system. *IEEE Trans. Magn.* **1999**, *35*, 4259–4267. [CrossRef]
6. Long, Z.; He, G.; Xue, S. Study of EDS & EMS hybrid suspension system with permanent-magnet halbach array. *IEEE Trans. Magn.* **2011**, *47*, 4717–4724. [CrossRef]
7. Hsu, Y.H.; Langhom, A.; Ketchen, D.; Holland, L.; Minto, D.; Doll, D. Magnetic levitation upgrade to the Holloman High Speed Test Track. *IEEE Trans. Appl. Supercond.* **2009**, *19*, 2074–2077. [CrossRef]
8. Okubo, T.; Ueda, N.; Ohashi, S. Effective control method of the active damper system against the multidirectional vibration in the superconducting magnetically levitated bogie. *IEEE Trans. Appl. Supercond.* **2016**, *26*. [CrossRef]
9. Lee, C.Y.; Jo, J.M.; Han, Y.J. Design, fabrication, and operating test of the Prototype HTS electromagnet for EMS-based maglev. *IEEE Trans. Appl. Supercond.* **2012**, *22*. [CrossRef]
10. Xu, J.; Geng, Q.; Li, Y.; Li, J. Design, fabrication and test of an HTS magnetic suspension experimental system. *IEEE Trans. Appl. Supercond.* **2016**, *26*. [CrossRef]
11. Yan, L. Suggestion for selection of Maglev option for Beijing-Shanghai high-speed line. *IEEE Trans. Appl. Supercond.* **2004**, *14*, 936–939. [CrossRef]
12. Du, J.; Ohsaki, H. Numerical analysis of eddy current in the EMS-Maglev system. In Proceedings of the 6th IEEE International Conference on Electrical Machines and Systems, 2003, ICEMS 2003, Beijing, China, 9–11 November 2003; pp. 761–764.
13. Ding, J.; Yang, X.; Long, Z.; Dang, N. Three dimensional numerical analysis and optimization of electromagnetic suspension system for 200 km/h maglev train considering eddy current effect. *IEEE Access.* **2018**, *6*, 1547–1555. [CrossRef]
14. Abdelrahman, A.S.; Sayeed, J.; Youssef, M.Z. Hyperloop transportation system: Analysis, design, control and implementation. *IEEE Trans. Ind. Electron.* **2018**, *65*, 7427–7436. [CrossRef]

15. Ji, W.Y.; Jeong, G.; Park, C.B.; Jo, I.H.; Lee, H.W. A study of non-symmetric double-sided linear induction motor for Hyperloop all-in-one system (Propulsion, Levitation, and Guidance). *IEEE Trans. Magn.* **2018**, *54*. [CrossRef]
16. Hao, L.; Huang, Z.; Dong, F.; Qiu, D.; Shen, B.; Jin, Z. Study on electrodynamic suspension system with high-temperature superconducting magnets for a high-speed maglev train. *IEEE Trans. Appl. Supercond.* **2019**, *29*, 1–5. [CrossRef]
17. Mulcahy, T.M.; He, J.; Rote, D.M.; Rossing, T.D. Forces on a magnet moving past figure-eight coils. *IEEE Trans. Magn.* **1993**, *29*, 2947–2949. [CrossRef]
18. Knowles, R. Dynamic circuit and Fourier series methods for moment calculation in electrodynamic repulsive magnetic levitation system. *IEEE Trans. Magn.* **1982**, *18*, 953–960. [CrossRef]
19. Lee, J.; Bae, D.K.; Kang, H.; Ahn, M.C.; Lee, Y.; Ko, T.K. Analysis on ground conductor shape and size effect to levitation force in static type EDS simulator. *IEEE Trans. Appl. Supercond.* **2010**, *20*, 896–899. [CrossRef]
20. Ko, W.; Ham, C. A novel approach to analyze the transient dynamics of an electrodynamic suspension maglev. *IEEE Trans. Magn.* **2007**, *43*, 2603–2605. [CrossRef]

© 2019 by the authors. Licensee MDPI, Basel, Switzerland. This article is an open access article distributed under the terms and conditions of the Creative Commons Attribution (CC BY) license (http://creativecommons.org/licenses/by/4.0/).

Article

Optimal Structural Design of a Magnetic Circuit for Vibration Harvesters Applicable in MEMS

Zoltán Szabó [1], Pavel Fiala [2,*], Jiří Zukal [3], Jamila Dědková [3] and Přemysl Dohnal [3]

[1] CVVOZE Centre—Department of Theoretical and Experimental Electrical Engineering, Brno University of Technology, Technicka 12, 616 00 Brno, Czech Republic; szaboz@feec.vutbr.cz
[2] SIX Centre—Department of Theoretical and Experimental Electrical Engineering, Brno University of Technology, Technicka 12, 616 00 Brno, Czech Republic
[3] Department of Theoretical and Experimental Electrical Engineering, Brno University of Technology, Technicka 12, 616 00 Brno, Czech Republic; xzukal03@stud.feec.vutbr.cz (J.Z.); dedkova@feec.vutbr.cz (J.D.); dohnalp@feec.vutbr.cz (P.D.)
* Correspondence: fialap@feec.vutbr.cz; Tel.: +420-604-076-280

Received: 28 November 2019; Accepted: 22 December 2019; Published: 6 January 2020

Abstract: The paper discusses the results of research into a vibration-powered milli- or micro generator (MG). The generator harvests mechanical energy at an optimum level, utilizing the vibration of its mechanical system. The central purpose of our report is to outline the parameters that are significant for implementing the actual design to harvest the maximum (optimum) energy possible within periodic symmetrical systems, while respecting the typical behavior of the output voltage. The relevant theoretical outcomes influence the measurability and evaluation of the physical quantities that characterize the designed structures. The given parameters, which are currently defined in millimeters, are also applicable within the micrometer range, or MEMS. The article compares some of the published microgenerator concepts and design versions by using effective power density, among other parameters, and it also brings complementary comments on the applied harvesting techniques. The authors demonstrate minor variations in the magnetic rotationally symmetric circuit geometry that affect the pattern of the device's instantaneous output voltage; in this context, the suitability of the individual design approaches that are to be used with MEMS as a vibration harvesting system is analyzed in terms of properties that are applicable in Industry 4.0.

Keywords: Harvesting; low-power applications; vibration; micro-generator; optimal solution; magnetic circuit; periodical structure; effective power density; symmetry

1. Introduction

In recent years, alternative sources of energy have become the main subject of numerous research projects [1–22], with the optimum energy conversion being one of the central points of focus. Such a transformation is often ensured through a vibration microgenerator [23–26]. Effective energy harvesters exploiting the mechanical vibrations and related non-stationary magnetic fields have already been investigated and reported [2].

The comparative approach applied to harvesters for milli- or micro generators (MGs) within study [18] allows for an effective evaluation of different conversion concepts, namely, interpretations of Faraday's law of induction. An optimal harvester design to yield the maximum power is obtainable via minor structural modifications that may substantially change the resulting performance while the parameters (including the weight, volume, and vibrations) remain virtually identical to those of standard harvesters. Different papers, including [2], detail functional magnetic circuits for vibration harvesters, where the model experiments and a comparison of various versions illustrate the effect of magnetic circuit modifications on the output voltage and power of a harvester. Advantageously, the devices can be grouped into periodic structures and also used in closed systems, such as automobiles, aircraft, and other units that are suitable for the inclusion of harvesters as additional and reliable energy sources. The possibilities of residual energy harvesting are examined in article [3]; the discussion comprises, among other aspects, specific harvester installation conditions, and requirements.

As regards the microgenerator design (Figure 1), Figure 2a introduces the most widely preferred principle (I, [2,3]); in the given context, it is necessary to respect the general conclusions of Faraday's law of induction as formulated in, for example, Equation (1) and Figure 2b below. Figure 2 shows multiple processes and elements, including the magnetization of the permanent magnet M; magnetic flux Φ; magnetic lines of force; oriented area S enclosed by the coil thread; electric coil; and, character of the generator's core motion with respect to the coil. The related Figure 2c then introduces a design version that minimizes the impact of external electromagnetic fields (non-stationary) on the principal function of the generator.

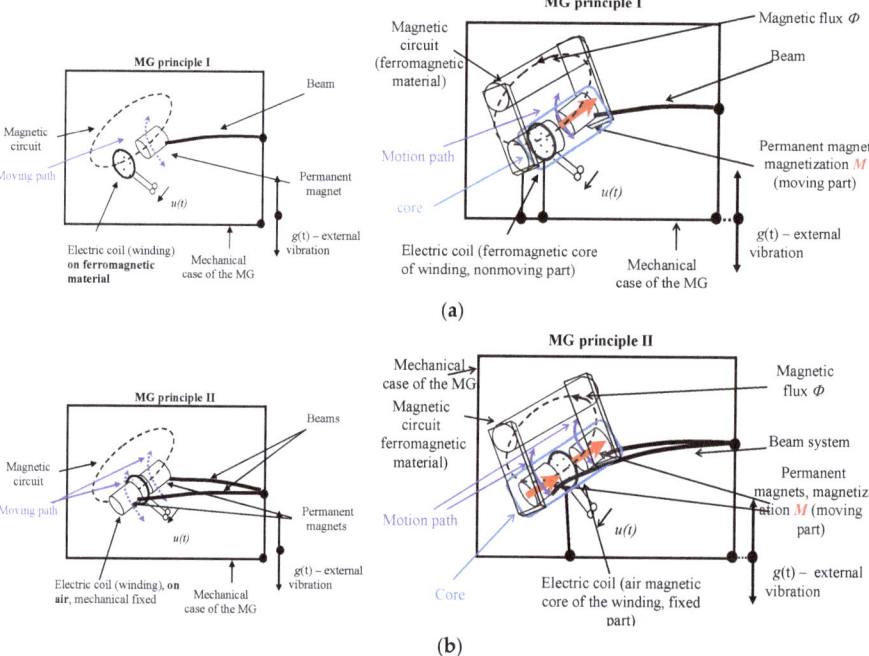

Figure 1. The principal configuration of the core of the milli- or micro generator (MG): (**a**) a beam version, principle I; (**b**) a beam version, principle II [2].

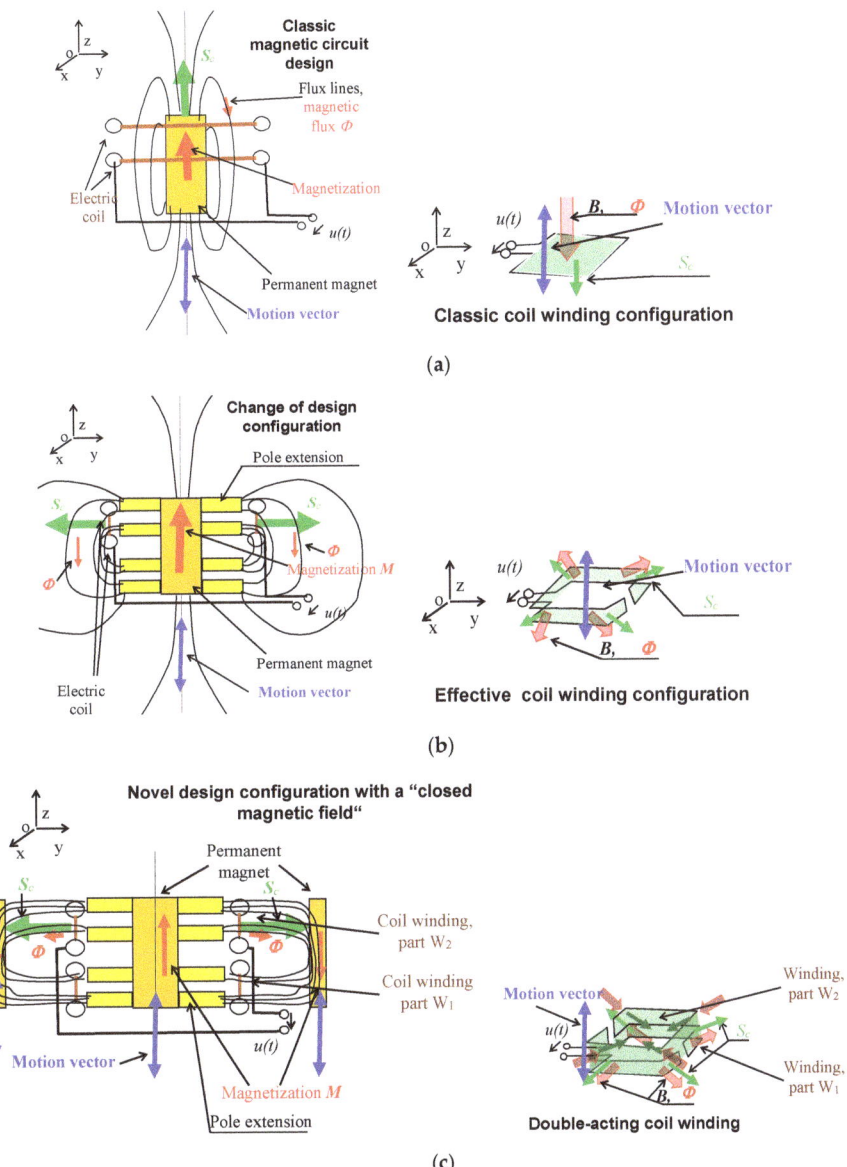

Figure 2. The basic arrangement of the investigated rotationally symmetric geometry device exploiting Faraday's induction law: (**a**) the classic solution; (**b**) the novel arrangement; and, (**c**) the option with a closed magnetic circuit to minimize (optimize) the impact of external magnetic fields [2].

The generator was modeled to facilitate optimal design of the dimensions (minimum size and weight m) [2]. The vibrations measured with critical positioning of the device reached the maximum of $G = 0.2$ g ($g = 9.81$ ms^{-2}). In the discussed concept, the resonance might vary, according to the origin of the vibrations, from the tuned resonance frequency f_r by up to tens of percent.

As regards the optimum design variant, the critical parameter consisted in the boundary sensitivity of the generator to the minimum vibration amplitude; the relevant value corresponded to 0.01 g–0.05 g.

In the following portions of the presentation, the proposed structural problems and methods for their solution will be discussed.

Microgenerator systems and relevant manufacturing methods were discussed on a comprehensive basis previously [2–5]; the structural details and consequences are indicated herein, as in Figures 2–4.

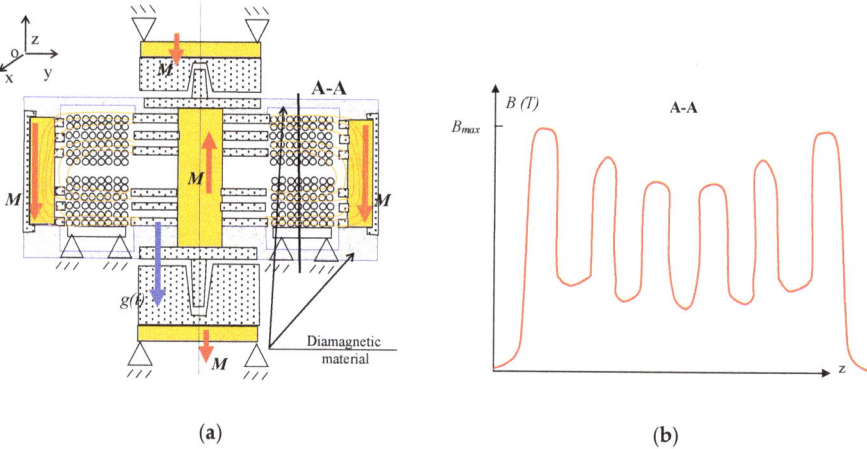

Figure 3. The basic arrangement of the investigated device exploiting Faraday's induction law: (**a**) the MG core; and (**b**) the magnetic flux density distribution along the z axis, line A-A.

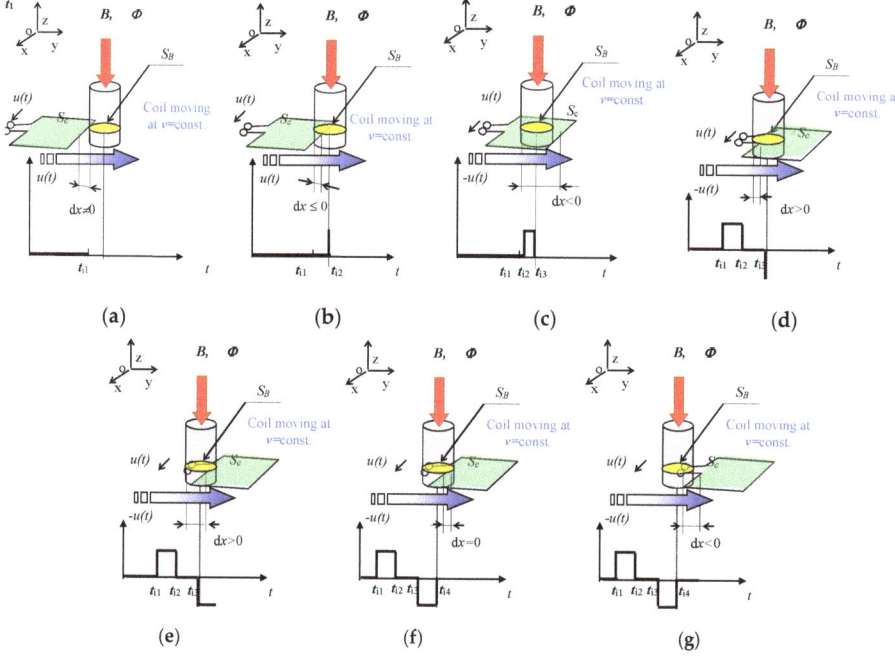

Figure 4. The electric voltage induction in the applied coil, (**a**–**g**), according to Faraday's law of induction [2].

2. Designing the MG

The microgenerator utilizes an external environment that is characterized by the occurrence of mechanical vibrations, exploiting a suitable mechanical coupling to dampen these vibrations and generate an electric power P_{out}. The required output power P_{out} of the optimal design depends on the type of the output load Z. The optimal arrangement of the MG is based on the concepts in Figure 1, Figure 2c, and Figure 3, with the magnetization orientation indicated. In terms of the mechanical properties, the device was discussed in dedicated papers and patents, such as [10,25]. Figure 4 presents details of the transformation process and electricity generation; the actual engineering approach adopted in solving these procedures then embodies the necessary precondition for the subsequent identification of the optimal design. The mathematical model outlined in [2] is, in a basic form, incorporated in the corresponding Formula (5), below.

Figure 3, as above, presents one of the progressive options available for seating the moving part of the generator, a solution that eliminates the classic spring or girded beam (Figure 1). The designed system (Figure 3), is tuned to the mechanical oscillation resonance frequency f_{res} and it constitutes the basis of the optimal approach. Such an arrangement allows for us to reach the maximum possible harvest rate and transform the field into an electric voltage; Figure 4 shows the corresponding preconditions.

3. Modeling the MG

To support our approaches, the paper includes fundamental parts of the relevant mathematical model, which is defined, for example, within referenced publications [1–3]. In the given context, the model can be formulated, as

$$\oint_\ell \mathbf{E}(t) \cdot d\ell = -\int_S \frac{\partial \mathbf{B}(t)}{\partial t} dS + \oint_\ell (\mathbf{v}(t) \times \mathbf{B}(t)) \cdot d\ell \tag{1}$$

where E(t) denotes the electric field intensity vector, B(t) is the magnetic flux density vector, v(t) represents the generator core position drift in time (the instantaneous velocity) vector, S stands for the cross section of the area with magnetic flux Φ, and l denotes the curve along the boundary of the S. Figure 4 illustrates the change of the magnetic flux of the field ($t_{i1}, \ldots, t_{i4}$) and also the resulting induction of the voltage u. The behavior of the voltage $u(t)$ can be evaluated by following the steps that are indicated in Figure 4; this behavior assumes the validity of Equation (1), magnetic flux Φ configuration, and electric coil shape with an active surface S_c.

We need to know the values of energy and transformation rate to be able to evaluate the efficiency of the proposed design (Figure 2c). The state equation can be defined with respect to the energy conservation law regarding the considered problem [1–3]. Subsequently, the kinetic and potential energies, W_k and W_p, respectively, which are related to the movement of the generator's core, can be defined as

$$W_k = \frac{1}{2} m v^2, \quad W_p = m g z. \tag{2}$$

where m is the mass of the MG system, v denotes the mean velocity, and g represents the gravity constant.

The equation of state used by the authors of [1] and [2] captures the electromechanical coupling in the device, being expressed as

$$m g z - \int_\ell \int_{V_J} (\mathbf{J} \times \mathbf{B}) dV \cdot n d\ell - \int_{V_{Jc}} \frac{1}{2} \frac{J^2}{\gamma} dV = \frac{1}{2} m \left(\frac{dz}{dt}\right)^2$$
$$\eta \int_{V_M} \frac{1}{2} \mathbf{B}_M \mathbf{H}_M dV = \frac{1}{2} m \left(\frac{dz}{dt}\right)^2 \tag{3}$$

where dz/dt is the moving part drift in time, further measurable as the velocity v; η represents the magnetic field transformation efficiency; $\mathbf{n}$ denotes the normal vector; γ is the specific conductivity of the wire; ℓ is the length of the shift caused by the specific strength; $\mathbf{B}$ denotes the magnetic flux density vector; $\mathbf{J}$ represents the current density vector; V_{J_c} stands for the coil wire volume; and, V_J is the volume of the electrically conductive components. The MG system then also includes the braking forces

$$m \frac{d^2 z}{dt^2} + l_c \frac{dz}{dt} + k\,z = F_z, \qquad (4)$$

where $d^2z/d\,t^2$ is the acceleration of the moving part, $dz/d\,t$ denotes the velocity v of the moving parts, m represents the mass, k stands for the stiffness coefficient, l_c is the damping coefficient, and F_z is the forces affecting the moving parts. The simplified model is described as

$$m \frac{d^2 z}{dt^2} + l_c \frac{dz}{dt}\,sign(\frac{dz}{dt})\,\frac{dz}{dt} + k\,z = (m_m + m_p)g(t) - \int_{V_J}\left(\left(\frac{dz}{dt}\times\mathbf{B}_{br}\right)\times\mathbf{B}\right)\cdot\mathbf{n}\,dV - \int_{\ell_{Jc}}(i\mathbf{n}\times\mathbf{B})\cdot\mathbf{n}\,d\ell, \quad (5)$$

where $\mathbf{B}_{br}$ is the braking magnetic flux density, J_v denotes the current density of the electrically conductive components, J_{circ} represents the current density in the coil winding, and i stands for the instantaneous value of the current through the coil. The geometrical model that is applied in ANSYS (Version 12, ANSYS inc., Houston, USA) is presented in sources [23], ref. [2] as well as Figures 3a and 5a. Figure 6, as below, shows the typical analysis of the ANSYS numerical model. The novel (optimal) generator design was tested on both a pneumatic and an electrodynamic shaker to verify the magnetic independence of the proposed solution. The magnetic circuit is designed such that its structure is enclosed within the body of the generator, ensuring reduced sensitivity to the external magnetic field and its changes. This parameter is of interest for application in the periodic structure of the outlined design usable in MEMS. The assumptions embodied in the variant from Figure 2c were experimentally verified.

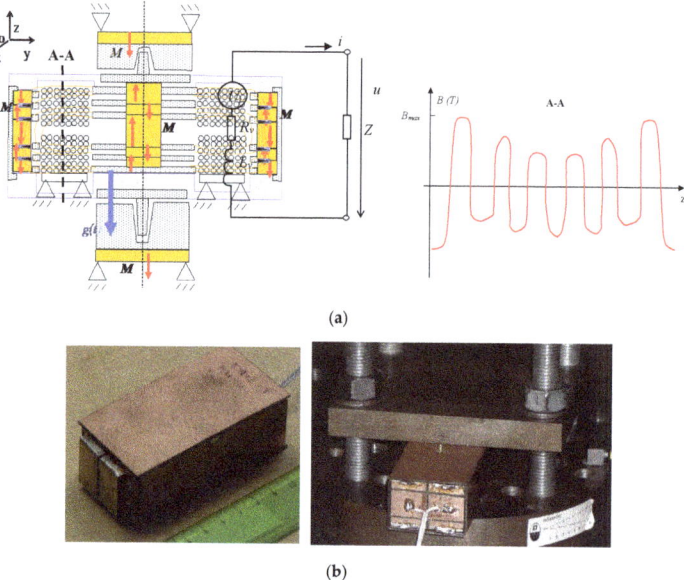

Figure 5. A geometrical model of the tested MG [2]: (**a**) the core of MG; and, (**b**) the functional sample subjected to a shaker-based test of the double-action winding.

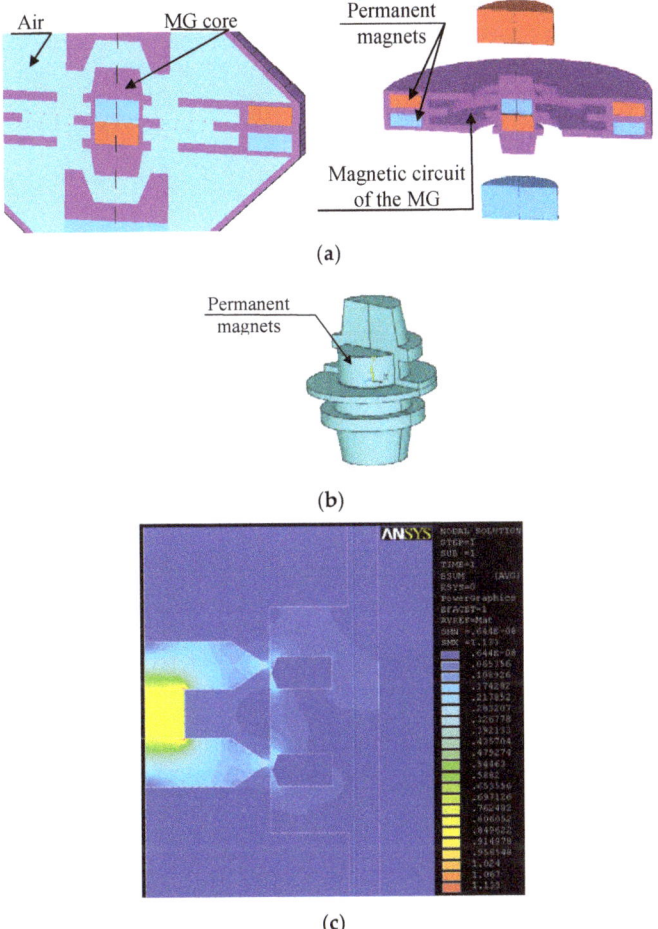

Figure 6. A geometrical rotationally SYMMETRIC model of the MG0 based on principle II [2]: (**a**) the ANSYS geometrical model; (**b**) the core; and, (**c**) the optimal design, detailed distribution of the magnetic flux density module B [T].

4. Selecting the MG Core Design

Within the design of the generator, the ANSYS system [4] was used for the numerical analysis and to optimize the key parts. A mathematical model exploiting partial differential equations further described the electromagnetic field distribution [1,2]. This model becomes evident from Formulas (3), (5); the non-linear equations, which define the behavior of the external electric circuit [1–3]; and, the mechanical model of the main parts of the generator. There is mutual action between the mechanical and the electromagnetic effects. The partial differential equations of a coupled electro-mechanical circuit [4,5] were used to build the physical model.

A simplified model was employed to design the generator components. The model utilized lumped parameters, as shown in Figure 7, and it also assumed various versions of the magnetic field changes described with respect to the phase perspective, as indicated in Figure 4a–g. The individual parts, comprising the permanent magnet, air gap, electrical winding, magnetic circuit, electric coil, one turn of the winding, pole extension, beam, body, and core of the magnetic part of the generator, are denoted by using the reference symbols in Figure 7a.

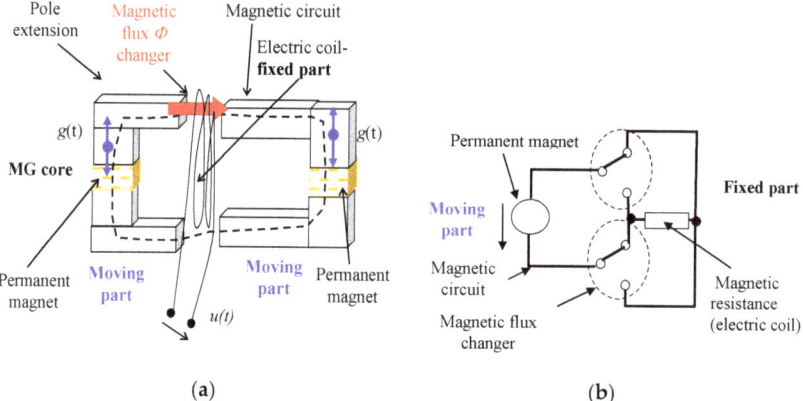

Figure 7. The MG core with a ferromagnetic circuit according to principle II [2]: (**a**) a structurally simple variant, acceleration g within the interval of 0.01 g–0.3 g; (**b**) a scheme of the configuration.

A lumped parameters model can describe these components in order to perform quick assessment of the properties of the proposed concept. The above structures are classifiable into two groups: One of these exploits the approach where a magnetic field moves with respect to a fixed coil, as shown in Figure 7b; the other then utilizes an electric coil moving with respect to a stationary magnetic field and fixed to the body of a generator. Regarding the above analysis, we also examine the concept of a magnetic field moving with respect to an electric coil fixed to the body of a generator based on principle II (Figure 3a), in which the ferromagnetic circuit is fully closed and its components do not move against each other.

The generator design versions were used to build a series of models in ANSYS and then employed to examine the vibration energy harvesting rate. As a result, we can demonstrate the distribution of the magnetic flux density module B and the magnetic field intensity module H in Figure 6c of the functional sample according to the configurations that are presented in Figure 6a,b.

5. Critical Parameters of the MG Design

The critical parameters are outlined in sources [1,2,26] and can be categorized into the following areas:

- mechanical dynamics;
- electromagnetic field; and,
- electronic systems (power management blocks).

In terms of the mechanical dynamics, the optimal state depends on finding an interval of the mechanical approach to a vibration system for the known resonant frequency f_{res}. Regarding this task, an aspect of major importance consists in the nonlinear stiffness coefficient k in the entire generator system (Figure 8). If the factor is adequately considered, then the device is capable of providing an operational efficiency of approximately 90%; in such conditions, the MG will operate at its maximum efficiency with minimal vibrations. The nonlinearity of the stiffness coefficient k depends on the choice of principal approach (Figure 1a,b and also Figures 6–8).

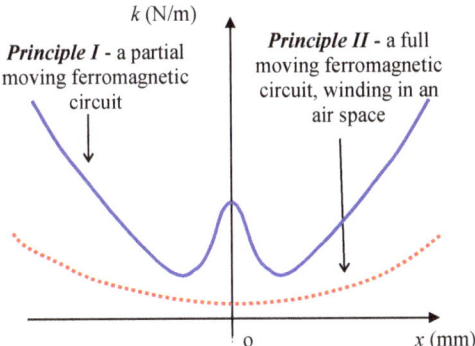

Figure 8. The applied stiffness characteristics [2], coefficient k; the behavior is nonlinear in both of the MG magnetic circuit principles.

Such an approach to the mechanical configuration of the MG is suitable for multiple purposes in microtechnology, including the formation of fields of resonant MGs (Figures 9–12). Two approaches were tested as regards the electromagnetic field: one utilizing air to substitute for the ferromagnetic material in the magnetic field (principle II, Figure 1b), and the other applying a ferromagnetic material according to Figure 3a. Figure 9 demonstrates a solution to facilitate the further development of the progressive concept (principle II) through changes of the dimensional parameters of the design, t_1, t_2, $\phi D_1 - \phi D_3$.

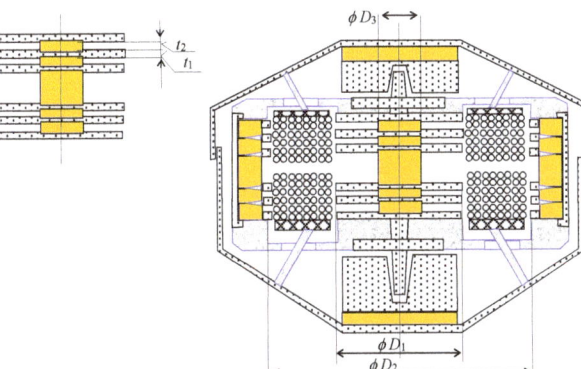

Figure 9. A component diagram of the MG0 design with the linearized coefficient of stiffness k (based on principle II).

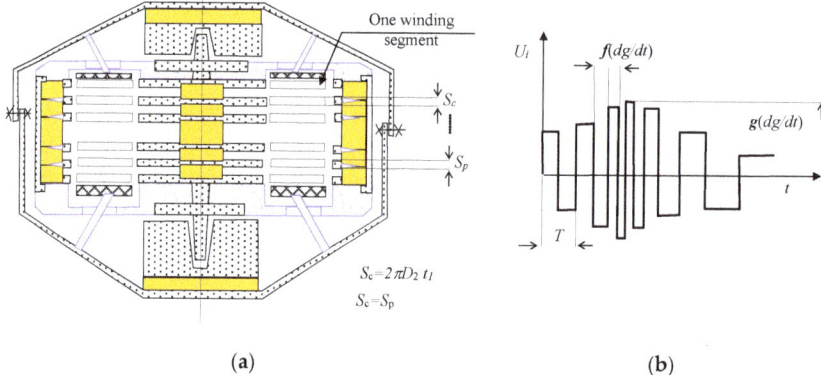

Figure 10. A component diagram of the MG0 design with the linearized coefficient of stiffness k (based on principle II); (**a**) configuration A, and (**b**) its output voltage U_{out}.

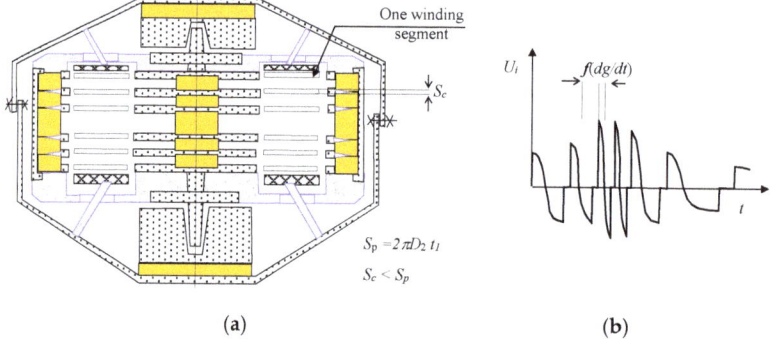

Figure 11. A component diagram of the MG0 design with the linearized coefficient of stiffness k (based on principle II); (**a**) configuration B, and (**b**) its output voltage.

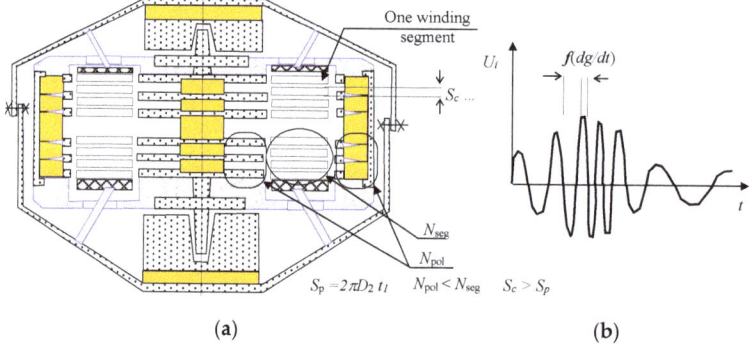

Figure 12. A component diagram of the MG0 design with the linearized coefficient of stiffness k (based on principle II); (**a**) configuration C, and (**b**) its output voltage.

Figures 10–12 demonstrate the difference in the shapes of the electric voltage induced in the coil of the generator at some characteristic settings of the design parameters t_1, t_2, $\phi D_1 - \phi D_3$.

6. Microstructures

In order to apply the above-defined principles and conclusions, we have to consider the relevant figures (MG principle II, Figures 1b, 3a and 9, Figures 10–12, where S_c is the effective area of the coil, Figure 2c; S_p denotes the area of the pole extension; t_1 represents the thickness of the pole extension of the MG core; N_{seg} is the number of the electrical winding segments; N_{pol} is the number of the pole extensions of the core and shell; and, f (dg/dt), g (dg/dt) denote the time variation of the gravitational acceleration of the moving part of the microgenerator. The correct setting of the MG0 structure, Figures 9–11, can be verified through measuring or evaluating the behavior of the output voltage on the terminals of the MG segment. The obtained instantaneous values of the patterns of the voltage $u(t)$ are then applicable in expressing, via the indirect method and based on the above formula (5), the observed physical quantities of the model.

As regards vibration energy harvesting within the microdimension, it is necessary to utilize in the MG fields the discussed principle II (Figure 7), together with certain variants of the relevant configurations of the magnetic circuit and winding (Figures 9–12). Thus, the preset requirements for the generator sensitivity and effective use of the space will enable us to harvest a high amount of residual energy. Figure 13, below, illustrates an exemplary periodic MG structure. The actual design (Figures 10–12) or other parameters can be altered to ensure the desired shape of the output voltage (Figures 10b, 11b and 12b) and also the conversion effectivity rate in transforming the mechanical vibrations to electrical energy.

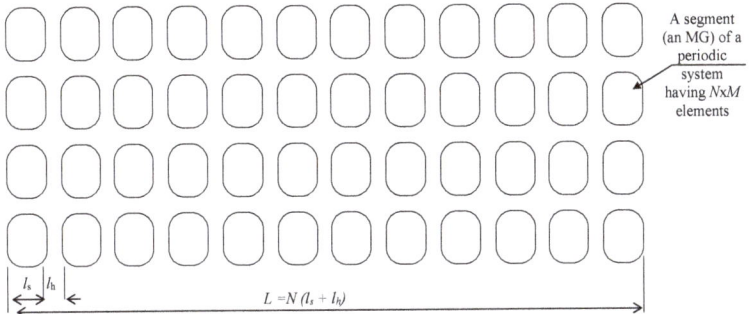

Figure 13. A field of periodically configured MGs an exemplary structure.

The optimal design of a symmetrically structured sensitive vibration harvester (principle II, Figure 1b to operate in the resonant band can be applied in segmentation into microgenerator structures, as in Figure 13). Segmented microgenerators clustered as units (a MEMS harvester) can be arranged into fields. In this type of configuration, the designer has to consider the condition in which the length of the excitation vibration wave is

$$\lambda_v \gg L, \; L = N(l_s + l_h) \tag{6}$$

where L denotes the length of a side of the periodic structure, l_s represents the length of the MG element, and l_h is the space between the elements of the periodically structured field of microgenerators. During the propagation of vibrations, the structure behaves such that the electric voltage is almost in phase at the output of the windings.

The actual engineering of the procedure to facilitate, especially in terms of the size, the transition from a minigenerator to a microdevice (MEMS) is accompanied by not only technological questions and problems, but also the fundamental requirement of respecting the principles that are characterized in this paper. Generally, it is possible to suggest that the set of usable magnetic materials comprises items such as nano Ni and convenient permalloys deposited via sputter coating or lamination. The weight of the flat structure, m_s, then determines the achievable resonant frequency, harvesting efficiency, and adjustment of the harvester's lower sensitivity limit. Importantly, each concrete application of the principle requires designing a suitable MEMS structure by using the above-shown models.

7. Comparing MG Concepts, Designs, and Structures

Current experiments with vibration microgenerators converting energy via magnetic induction (vibration/electric energy) employ various harvesting approaches (principle I, principle II) and magnetic circuit structures [2]; thus, the devices exhibit diverse output power and conversion efficiency rates with respect to the given size and vibration frequency spectrum [12–22]. When engineering a periodic microstructure, a designer has to consider the degree of efficiency at which the transformed energy (mechanical vibrations) is to be harvested, and they then select the microstructure element accordingly, while utilizing available technologies. Several specific methods and the obtained results by different research groups are compared below, Figures 14–18; in this context, the relevant concepts and structures of vibration generator transformation elements applied internationally are also discussed in view of the samples MG I–MG IV presented herein (Figures 19–21).

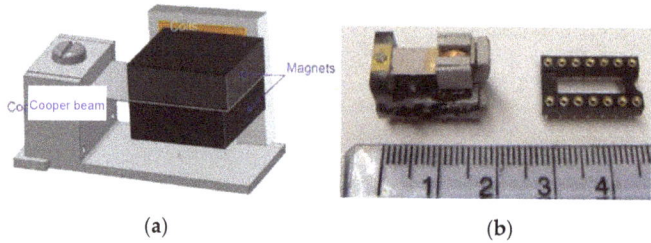

Figure 14. The silicon-based concepts developed by (**a**) Kulkarni et al. [11] and (**b**) Zhu et al. [13].

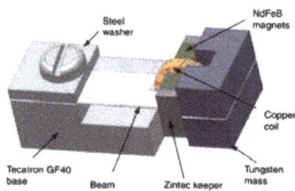

Figure 15. The vibration microgenerator designed by Beeby et al. [12].

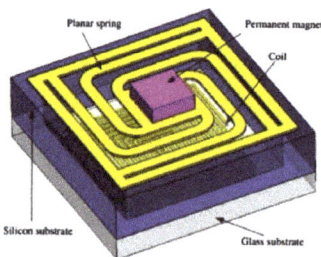

Figure 16. The electromagnetic design by Wang et al. [15].

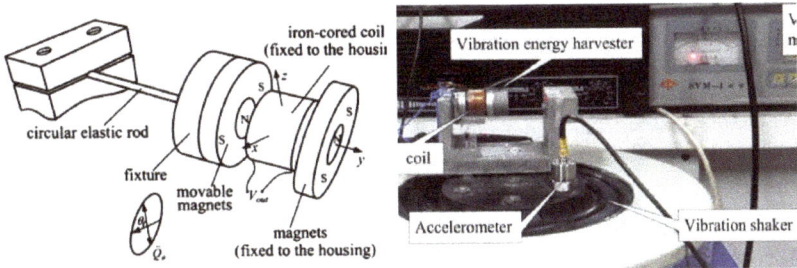

Figure 17. The electromagnetic vibration energy harvester using cylindrical geometry, developed by Yang et al. [16].

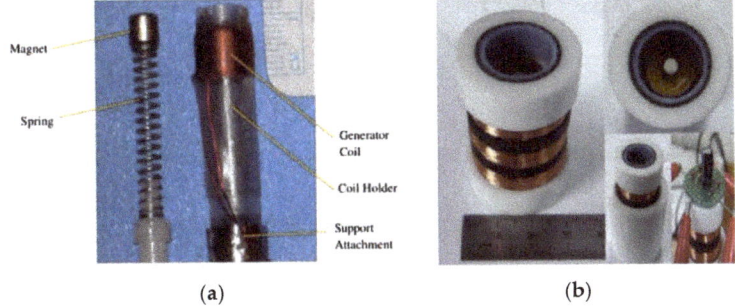

Figure 18. The permanent magnet (**a**) having a spring in the cylindrical structure of the energy harvester [18], (**b**) tested sample designed by Lee et al. [17].

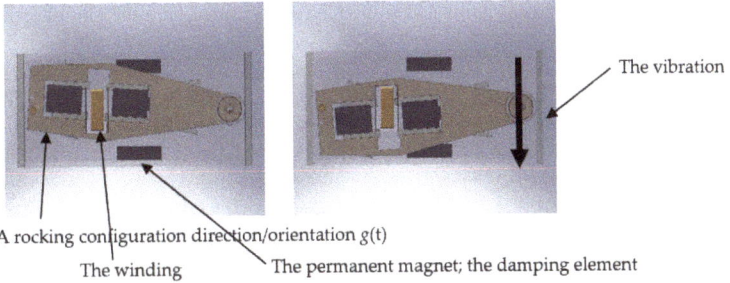

Figure 19. The basic symmetric magnetic circuit, structural design [2] according to Principles I and II, invariably with magnetic damping in the limit position of the rocking arm [3].

Figure 20. The tested microgenerator [2] based on Principle I: (**a**) MG I—the dimensions of 90 × 40 × 30 mm, $U_{out\,max}$ = 300 V; (**b**) MG II—the dimensions of 50 × 27 × 25 mm, $U_{out\,max}$ = 20 V; and (**c**) the instantaneous behavior of the output electrical voltage in MG I and MG II (the effect of the stiffness coefficient k—Figure 8).

Figure 21. The devices based on Principle II [2]: (**a**) MG III—the dimensions of 50 × 25 × 25 mm, $U_{out\,max}$ = 10 V; (**b**) MG IV—the dimesnions of 50 × 35 × 25 mm, $U_{out\,max}$ = 20 V; and, (**c**) the instantaneous behavior of the output electrical voltage in MG III and MG IV (the effect of the stiffness coefficient k—Figure 8).

The microgenerators that are characterized in Figures 14–18 correspond to the concepts and design versions of the vibration generator magnetic circuit and housing outlined by the authors of this paper. The solution from Figure 2a corresponds to the embodiments that are discussed within articles [14,15], comprising a clearly open magnetic circuit and an induction coil unfavorably positioned with respect to the movement of the permanent magnet. The concepts and tests that are presented in [11,13] relate to the configuration from Figure 2b, where the induction coil is positioned and oriented such that the harvester provides a higher efficiency; however, the magnetic circuit is not markedly closed. The technique adopted by Yang et al. [16] approaches the effective configuration from Figure 2c; the researchers employed the non-linear, non-monotonous function of the stiffness coefficient k, namely, the function specified as the solution respecting Principle I, Figure 1a). By contrast, Beeby et al. [12] proposed an interpretation that, when compared to our investigation, resembles the system stiffness coefficient within Principle II–Figure 1b.

At the DTEEE FEEC, BUT, comparative tests were performed of a vibration generator (Figure 19) with magnetic damping [2] of the mobile arm's movement; these testing cycles comprised design variants MG I–MG II according to Principle I and also versions MG III–MG IV exploiting Principle II. The parameters obtained in selected generators are summarized in Table 1; a wider comparison is available in study [18] (as in table 4).

Table 1. The parameters of selected vibration generators.

Reference	Permanent Magnet Type	Generator Body Size x, y, z [m]	Resonant Frequency f_r [Hz]	Amplitude Mech. Part A [m]	Output Power P_{out} [W]	Output Voltage U_{out} [V]	Load Resistance R [Ω]	Acceleration G, g = 9.81 [m/s]	Effective Power Density [W/m³]
Beeby et al. [12], 2007	–	375 mm³	52	–	2×10^{-6}	0.428 RMS	4000	0.06 g	≈6
Zhu et al. [13], 2010	FeNdB	2000 mm³	67.6–98	0.6×10^{-3}	$61.6–156.6 \times 10^{-6}$	–	–	0.06 g	≈30–80
Kulkarni et al. [11], 2008	FeNdB	3375 mm³	60–9840	1.5×10^{-3}	0.6×10^{-6}	0.025	52,700	0.398–4 g	≈0.2
Wang et al. [15], 2007	FeNdB	256 mm³	121.25	0.738×10^{-3}	–	0.06	–	1.5 g	–
Lee et al. [17], 2012	FeNdB	1.4×10^{-4} m³	16	–	1.52×10^{-3}	4.8	5460	0.2 g	≈10
Yang et al., [16], 2014.	–	50,000 mm³	22–25	–	13.4×10^{-3}	0.7–2.0	110	0.6 g	≈270
Elvin et al., [14], 2011	–	15,000 mm³	112	–	4×10^{-6}	0.007	986	–	≈0.26
MG I [2], 2006	FeNdB	90, 40, 30 mm	20–35	$50 \times 10^{-6}–400 \times 10^{-6}$	70×10^{-3}	4–60 (300) P-P	7500	0.15–0.4 g	≈650
MG II [2], 2006	FeNdB	50, 27, 25 mm	17–25	$50 \times 10^{-6}–400 \times 10^{-6}$	19.5×10^{-3}	6–15	5000	0.1–0.7 g	≈60
MG III	FeNdB	50, 25, 25 mm	21–31.5	$50 \times 10^{-6}–400 \times 10^{-6}$	5.0×10^{-3}	1.0–2.5	600	0.05–0.4 g	≈15
MG IV	FeNdB	50, 35, 25 mm	21–31.5	$50 \times 10^{-6}–400 \times 10^{-6}$	8.0×10^{-3}	1.0–2.5	1200	0.05–0.4 g	≈18
*Lith. battery [19], 2018									≈40 × 10⁶
*supercap [20], 2010									≈3–5
*fuel									≈4 × 10⁹
*U₂₃₅									≈9 × 10¹⁶

If application in microelectronics and periodic systems is assumed, then the solution displayed in Figure 2c appears to be advantageous; however, at major vibrations, namely, ones between 0.3 g and 1.0 g, it is beneficial to configure the magnetic circuit and coil as set out within Principle I, Figure 1a). Where the external vibrations drop below 0.3 g (0.01 g–0.2 g), Principle II, Figure 1b, has to be applied. A generator configuration design requires an analysis of the magnetic field expected for the active section of the device and overall presetting of the maximum values of the specific magnetic flux density B into the air gap that is to contain the generator winding, Figure 6; such an analysis and presetting have to facilitate the maximum magnetic flux change in time and space, as formulated within Faraday's law of induction (1) and to enable geometrical configuration of the winding shown in Figure 4. In the given context, it is advantageous to employ the double action system to facilitate a magnetic flux change, as indicated in Figures 5a and 7.

Figure 20 presents the tested generators MG I–MG II that exploit Principle I and the energy harvesting efficiency rates yielded from Figure 2c, exhibiting various size versions together with different parameters and measured patterns of the output electrical voltage U_{out}.

Figure 21 presents the embodiments of MG III and MG IV respecting the magnetic circuit configuration according to Principle II and the energy harvesting effectivity scheme that characterizes the variant from Figure 2c.

The last column of Table 1 comprises data that are related to effective power density [W/m^3]; this quantity enables us to express the effectivity of individual generator concepts and structures as regards harvesting quality. The last four lines indicate a comparable quantity for fossil and nuclear fuels, batteries, and supercaps.

For comparison purposes, Table 1 contains a quantity denoted as "effective power density"; in this context, it would probably be interesting to also indicate the harvesting rate of the microgenerator, but such a task appears to be rather problematic. Although the efficiency of a resonant harvester can be preset to a desired level, as demonstrated via Formulas (1)–(5), the associated model (5), and the test cycles visualized in Figures 19–21 the achievable efficiency rate markedly depends on the quality of the mechanical coupling between the vibration source, the power management unit, and other relevant parameters; our tests yielded final rates between 50 and 95%. The problem was analyzed by different authors already previously [2–5,24].

8. Conclusions

The paper discussed the outcomes of a theoretical investigation into the design and principles of mini/micro generators to facilitate mechanical vibration energy harvesting [2]. The main product of the continuous research consists in simulation-based determination of the optimum rotationally symmetric geometry design versions and parameters of a relevant magnetic circuit.

The advantageous solutions and options are embodied in the generator design versions according to the proposed principles I and II, which ensure the necessary resistive loads and associated impedances. Exploiting the measured output voltages of the selected variants, the derived theoretical models can evaluate the harvester quality and fabrication procedure. Using the hybrid measurement approach combined with a numerical model, it is possible to classify other physical quantities of the electromagnetic field inside the generator.

The harvesters fabricated according to principle II, utilized in the range of f_r = 10–50 Hz (frequent in the automotive and aeronautics sectors), are integrable into miniaturized microgenerator structures working within the range of G = 0.05 g–0.08 g. This concept could advantageously employ in practice the higher level of vibrations available compared to the design based on principle I [3]. The generators that employ principle I operate at vibration levels higher than $G \geq 0.15$ g. Generally, the winding configuration variants convenient for the frequency ranges of f_v = 1–10 Hz, f_v = 50–150 Hz are demonstrated in Figure 2b,c.

We discussed selected samples of microgenerators to evaluate multiple quantities, including the effective power density (Table 1). This quantity is utilized as the parameter enabling us to choose the source of energy applicable in the given task or unit and determine whether the actual selection of the correct approach is a parameter for facilitating effective designing of MEMS structures. The knowledge that was obtained through the experiments is beneficial for the use of autonomous electro-mechanico-electronic systems in Industry 4.0 projects.

Author Contributions: P.F. contributed to the theoretical part, numerical model, and experiments; he also wrote the paper. Z.S. conceived and designed the experiments; J.D. contributed to the optimization procedures; and P.D., J.Z. evaluated the experiments and graphics. All authors have read and agreed to the published version of the manuscript.

Acknowledgments: The research was financed by the National Sustainability Program under grant No. LO1401 and supported within a grant of Czech Science Foundation (GA 17-00607S). For the actual analyses and experiments, the infrastructure of the SIX Center was used.

Conflicts of Interest: The authors declare no conflict of interest. The founding sponsors had no role in the design of the study; in the collection, analyses, or interpretation of data; in the writing of the manuscript, and in the decision to publish the results.

References

1. Stratton, J.A. *Theory of Electromagnetic Field*; Czech Version; SNTL: Praha, Czech Republic, 1961.
2. Szabo, Z.; Fiala, P.; Dohnal, P. Magnetic circuit modifications in resonant vibration harvesters. *Mech. Syst. Signal Process.* **2018**, *99*, 832–845. [CrossRef]
3. Fiala, P. *Vibration Generator Conception*; Research Report in Czech; DTEEE, Brno University of Technology: Brno, Czech Republic, 2003; p. 95.
4. Jirků, T.; Fiala, P.; Kluge, M. Magnetic resonant harvesters and power management circuit for magnetic resonant harvesters. *Microsyst. Technol.* **2010**, *16*, 677–690. [CrossRef]
5. Fiala, P.; Jirků, T. Analysis and Design of a Minigenerator, Progress in Electromagnetics 2008. In Proceedings of the PIERS 2014 in Guangzhou, Guangzhou, China, 25–28 August 2014; pp. 749–753.
6. Madinei, H.; Khodaparast, H.H.; Adhikari, S.; Friswell, M. Design of MEMS piezoelectric harvesters with electrostatically adjustable resonance frequency. *Mech. Syst. Signal Process.* **2016**, *81*, 360–374. [CrossRef]
7. Scapolan, M.; Tehrani, M.G.; Bonisoli, E. Energy harvesting using parametric resonant system due to time-varying damping. *Mech. Syst. Signal Process.* **2016**, *79*, 149–165. [CrossRef]
8. Gatti, G.; Brennan, M.; Tehrani, M.; Thompson, D. Harvesting energy from the vibration of a passing train using a single-degree-of-freedom oscillator. *Mech. Syst. Signal Process.* **2016**, *66*, 785–792. [CrossRef]
9. Davino, D.; Krejčí, P.; Pimenov, A.; Rachinskii, D.; Visone, C. Analysis of an operator-differential model for magnetostrictive energy harvesting. *Commun. Nonlinear Sci. Numer. Simul.* **2016**, *39*, 504–519. [CrossRef]
10. Saravanan, S.; Dubey, R. Optical absorption enhancement in 40 nm ultrathin film silicon solar cells assisted by photonic and plasmonic modes. *Opt. Commun.* **2016**, *377*, 65–69. [CrossRef]
11. Kulkarni, S.; Koukharenko, E.; Torah, R.; Tudor, J.; Beeby, S.; O'Donnell, T.; Roy, S.; Beeby, S. Design, fabrication and test of integrated micro-scale vibration-based electromagnetic generator. *Sens. Actuators A Phys.* **2008**, *145*, 336–342. [CrossRef]
12. Beeby, S.P.; Torah, R.N.; Torah, M.J.; O'Donnell, T.; Saha, C.R.; Roy, S. A microelectromagnetic generator for vibration energy harvesting. *J. Micromech. Microeng.* **2007**, *17*, 1257–1265. [CrossRef]
13. Zhu, D.; Roberts, S.; Tudor, M.J.; Beeby, S.P. Design and experimental characterization of a tunable vibration-based electromagnetic micro-generator. *Sens. Actuators A Phys.* **2010**, *158*, 284–293. [CrossRef]
14. Elvin, N.G.; Elvin, A.A. An experimentally validated electromagnetic energy harvester. *J. Sound Vib.* **2011**, *330*, 2314–2324. [CrossRef]
15. Wang, P.-H.; Dai, X.-H.; Fang, D.-M.; Zhao, X.-L. Design, fabrication and performance of a new vibration-based electromagnetic micro power generator. *Microelectron. J.* **2007**, *38*, 1175–1180. [CrossRef]
16. Yang, J.; Yu, Q.; Zhao, J.; Zhao, N.; Wen, Y.; Li, P.; Qiu, J. Design and optimization of a bi-axial vibration-driven electromagnetic generator. *J. Appl. Phys.* **2014**, *116*, 114506. [CrossRef]
17. Lee, B.-C.; Rahman, A.; Hyun, S.-H.; Chung, G.-S. Low frequency driven electromagnetic energy harvester for self-powered system. *Smart Mater. Struct.* **2012**, *21*, 125024. [CrossRef]
18. Siddique, A.R.M.; Mahmud, S.; Van Heyst, B. A comprehensive review on vibration based micro power generators using electromagnetic and piezoelectric transducer mechanisms. *Energy Convers. Manag.* **2015**, *106*, 728–747. [CrossRef]
19. Quinn, J.B.; Waldmann, T.; Richter, K.; Kasper, M.; Wohlfahrt-Mehrens, M. Energy Density of Cylindrical Li-Ion Cells: A Comparison of Commercial 18650 to the 21700 Cells. *J. Electrochem. Soc.* **2018**, *165*, A3284–A3291. [CrossRef]
20. Liu, C.; Yu, Z.; Neff, D.; Zhamu, A.; Jang, B.Z. Graphene-Based Supercapacitor with an Ultrahigh Energy Density. *Nano Lett.* **2010**, *10*, 4863–4868. [CrossRef] [PubMed]
21. Zhu, D.; Tudor, M.J.; Beeby, S.P. Strategies for increasing the operating frequency range of vibration energy harvesters. *J. Meas. Sci. Technol.* **2010**, *21*, 22001. [CrossRef]
22. El-Hami, M.; Glynne-Jones, P.; White, N.; Hill, M.; Beeby, S.; James, E.; Brown, A.; Ross, J. Design and fabrication of a new vibration-based electromechanical power generator. *Sens. Actuators A Phys.* **2001**, *92*, 335–342. [CrossRef]
23. ANSYS Users Manual, (1991–2019). USA. Available online: www.ansys.com (accessed on 1 July 2018).
24. Fiala, P.; Hadas, Z.; Singule, Z.; Ondrusek, C.; Szabo, Z. A Vibrational Generator for Electric Energy Producing, (in CZ Vibrační Generator pro Výrobu Elektrické Energie). Patent No. 305591-CZ, 12 September 2006.

25. Fiala, P. Fotovoltaický Element Zahrnující Rezonátor (A Photovoltaic Element with an Included Resonator). Patent No. 303866, 27 January 2011.
26. Fiala, P.; Szabo, Z.; Marcon, P.; Roubal, Z. Mini-and Microgenerators Applicable in the MEMS Technology, Smart Sensors, Actuators, and MEMS VIII. *Proc. SPIE* **2017**, *10246*, 1–8. [CrossRef]

© 2020 by the authors. Licensee MDPI, Basel, Switzerland. This article is an open access article distributed under the terms and conditions of the Creative Commons Attribution (CC BY) license (http://creativecommons.org/licenses/by/4.0/).

MDPI
St. Alban-Anlage 66
4052 Basel
Switzerland
Tel. +41 61 683 77 34
Fax +41 61 302 89 18
www.mdpi.com

Symmetry Editorial Office
E-mail: symmetry@mdpi.com
www.mdpi.com/journal/symmetry

www.ingramcontent.com/pod-product-compliance
Lightning Source LLC
LaVergne TN
LVHW070712100526
838202LV00013B/1077